The Parasites of *Homo sapiens*

Also available from Taylor & Francis

Malaria in Pregnancy
Deadly parasite, susceptible host
Edited by Patrick E. Duffy and Michael Fried
ISBN 0-415-27218-1 (hardback)

Malaria
Molecular and clinical aspects
Edited by Mats Wahlgren and Peter Perlmann
ISBN 90-5702-446-2 (hardback)

Parasites of the Colder Climates
Edited by Hannah Akuffo, Ewert Linder, Inger Ljunström and Mats Wahlgren
ISBN 0-415-27584-9 (hardback)

Interrelationships of the Platyhelminthes
Edited by D. T. J. Littlewood and R. A. Bray
ISBN 0-7484-0903-3 (hardback)

The Biology of Nematodes
Edited by Donald L. Lee
ISBN 0-415-27211-7 (hardback)

Parasitology
Jack Chernin
ISBN 0-7484-0817-7 (paperback)

Handbook of Drugs for Tropical Parasitic Infections
Yakoub Aden Abdi, Lars L. Gustafsson, Orjan Eriksson, Urban Hellgren
ISBN 0-7484-0167-9 (hardback)

Bird Haemosporidian Parasites
Edited by Gediminas Valkiunas
ISBN 0-415-30097-5 (hardback)

The Parasites of *Homo sapiens*

An annotated checklist of the Protozoa, Helminths and Arthropods for which we are home

R W Ashford and W Crewe

Second edition

LONDON AND NEW YORK

This edition first published 2003 by Taylor & Francis
11 New Fetter Lane, London EC4P 4EE

Simultaneously published in the USA and Canada
by Taylor & Francis Inc.
29 West 35th Street, New York, NY 10001

Taylor & Francis is an imprint of the Taylor & Francis Group

Prepared from electronic files supplied by the authors
Printed and bound in Great Britain by St Edmundsbury Press,
Bury St Edmunds, Suffolk

Every effort has been made to ensure that the advice and information in this book is true and accurate at the time of going to press. However, neither the publisher nor the authors can accept any legal responsibility or liability for any errors or omissions that may be made. In the case of drug administration, any medical procedure or the use of technical equipment mentioned within this book, you are strongly advised to consult the manufacturer's guidelines.

British Library Cataloguing in Publication Data
A catalogue record for this book is available from the British Library

Library of Congress Cataloguing in Publication Data
Ashford, R. W.
The Parasites of *Homo sapiens*: an annotated checklist of the protozoa, helminths and arthropods for which we are home / R. W. Ashford and W. Crewe
p. cm.
Previous ed. published: Liverpool: Liverpool School of Tropical Medicine, 1998
Includes bibliographical references and index.
1. Medical parasitology – Handbooks, manuals, etc. I. Crewe, W. II. Title.

QR251.A83 2003
616.9'6 – dc21 200231961

ISBN 0-415-27688-8 (pbk)
ISBN 0-415-31118-7 (hbk)

PREFACE

In 1998, we presented a simple but comprehensive checklist of all those animals which find a home in or on the human body. To our knowledge, such a list had not previously been published. This was somewhat surprising as we are surely one of the most parasitised of all species and our parasite fauna can be used to illustrate and develop numerous generalisations on parasitic infection in mammals as a whole. The checklist and annotations were intended to be useful not only to those interested in the practical aspects of disease control but also to zoologists and parasitologists with more theoretical interests. This second edition closely follows the pattern of the first, but incorporates much new information that had been missed, or has been more recently published.

A large proportion of the information compiled here is extracted from the meticulous works of the late Professor Paul Beaver and his colleagues and from that of Dr Isabel Coombs and Professor David Crompton. Without these invaluable precursors it is doubtful that this work would have been possible. We are particularly grateful to Professor Mike Service, who helped with the arthropods, and to Professor George Nelson, who kindly reviewed the many groups of which he has personal knowledge. Professor Silvio Pampiglione and Dr Odile Bain assisted with the filarial worms, Professor Phil Craig with the cestodes, and Dr Marie-Claude Durette-Desset with the trichostrongyles. Dr Kosuke Haruki kindly assisted with access to some of the Japanese literature, and Dr Patrick Giraudoux commented usefully on the introduction. For this second edition, Dr Ralph Muller's new edition of *Worms and Human Disease* has been very useful; comments have been kindly received from Professor R.C. Anderson, Professor Stanley Dissanaike, one parasitologist who wishes to be anonymous, Dr Alan Curry, Dr Ralph Muller, Professor N. Chowdhury, Professor John Cross and Professor J.-Y. Chai. Dr David Gibson kindly allowed us to preview part of his forthcoming revision of the Trematoda, from which the order of the families and species has been taken. Drs Gedimanas Valkiunas and Tatjana Iezhova kindly scoured all 25 volumes of Skrjabin's *Essentials of Trematodology*, looking for synonymies. The reorganisation of the Protozoa closely follows a manuscript kindly made available by Professor Frank Cox.

The errors and inadequacies that inevitably remain are, of course, our own responsibility.

The first edition was extensively copy-edited and prepared in camera-ready copy by Dr Keith Wallbanks. We are most grateful for the time, patience and expertise that he contributed to this task.

Contents

Introduction

Homo sapiens ranks among the most parasitised of all animals. In part this is because we know so much about all aspects of the biology of our species but, in addition, our varied habitat and diet and our global distribution must expose us to more infections than any other species. There must be few parasitic species which have never had the opportunity to infect a human.

Whereas some parasitic infections are responsible for much human disease and suffering, and are familiar to us all, the great majority are rare or obscure forms ignored by all but the most comprehensive texts. Surprisingly, there seems to have been no previous attempt to produce a comprehensive list of them all. The original purpose of this catalogue was purely personal; it originated as the data source for a largely speculative article on the origins of our parasite fauna and the lessons this might teach us (Ashford 1991, 2000). The growing interest in obscure infections which occasionally 'emerge', causing frightening outbreaks, and the response to the first edition of this book, have led us to believe that a full list of our parasites, with an indication of their distribution, abundance and zoonotic status, would be useful to a wider readership. We have therefore tried to produce as full as possible a list of the parasites that have been recorded as infecting *H. sapiens* under natural conditions.

Since the main purpose was to have a list which could be analysed in terms of possible patterns in the origins of human parasites, the notes which accompany entries deal mainly with taxonomic details or with information relevant to speculation on origins.

The term **parasite** is a Humpty-Dumpty-ism: it is used to mean many different things, so should be defined in each instance. We use it to mean a eukaryote (animal) organism for which another organism (the host) can be described as its habitat, which has not been shown to benefit the host. Note that this definition applies to **individual** organisms, at a given (extended) **time**, not to species as a whole. It is, for example, wrong to say that *Fasciola hepatica* as a species is a parasite: the adults are, and so is one of the larval stages, but there are two free-living larval stages as well. The catalogue includes all those species which depend on at least one parasitic stage in the life-cycle (obligate parasites), as well as normally free-living forms in which one or more stages are sometimes parasitic (facultative parasites).

Organisms for which *H. sapiens* is a paratenic host (in which the parasite survives but does not grow or develop) are included, although these do not gain nutritional benefit.

Some definitions of parasitism include harmfulness as an essential component. We do not agree with this, and have included many organisms that are apparently harmless. The notion of harm is very difficult to define, let alone substantiate. For example, an infection with 1000 hookworms is certainly harmful, but to say that a single hookworm is one-thousandth as harmful is as silly as to say that a glass of water is harmful because 1000 glasses would kill you. It may even be that some parasitic infections are in fact beneficial so long as the numbers are low: the effect of a few hookworms may be very like that of aspirin in preventing heart disease! Further, an infection that is harmful to **individuals** may benefit a host **population**. The malarial parasites largely protected the human population of West Africa from colonisation by Europeans. This phenomenon of protection of host populations against competition, by their parasites, has been termed the 'Wells effect' (Ashford 1997).

Ticks and trombiculid mites have been excluded from the checklist, although some of these remain on their hosts for relatively long periods of time and might therefore come within our definition of parasite. To have attempted to include all members of these groups which have been recorded feeding on man would have been impractical and would have led to an unbalanced record.

Even the term 'animal' has become doubtful: recent studies indicate that the Microsporidia are closer to fungi than to 'protozoa', and fungi are closer to metazoan animals than are the traditional protozoa. *Blastocystis* is also no longer a protozoan, but a 'stramenopyle' related, surprisingly, to *Opalina* and various algae. We have continued to exclude the traditional fungi, and to include traditional protozoan groups.

In other aspects also, the list is far from perfect. In many instances there are taxonomic questions which limit the interpretation of host specificity. There are numerous examples of morphologically similar organisms, infecting man and other hosts, where it is not clear whether one or two species are involved. Occasionally, investigation by biochemical and molecular–genetic methods has provided useful answers, as with *Entamoeba*, *Leishmania*, and *Ascaris* species, but even the most profound study sometimes fails to provide clear answers, as with *Giardia*. There remain numerous examples for which taxonomic study is still required before definite statements can be made about their host specificity. We hope that this work will encourage the elucidation of some of these questions.

Nomenclature is always a problem. We have tried to identify synonyms, and to use recently accepted names, but no-one will be happy with all our decisions. As this is not a taxonomic work, only widely used synonyms have been included, and we have not felt it useful to include the names of the authors of each species.

The reliability of published records is also a source of error. For example, the inclusion of one species in the first edition was found to have been based on pure invention. A specimen sent to Liverpool for confirmation, prior to publication as the first recorded human infection with *Diphyllobothrium* sp. in a particular country, further illustrates the problem: the specimen was, in fact, a gherkin. Fortunately this extreme example was not published, but there is little doubt that some of the records in the literature are equally spurious.

This original checklist was largely based on the works of Beaver *et al.* (1984) and Coombs and Crompton (1991), which provide a comprehensive foundation. The work of Beaver *et al.* (1984) is now a little out of date and does not express the explicit intention to be comprehensive. Coombs and Crompton (1991) covered only the helminths, for which they provide a comprehensive list. They include many records of pseudoparasitism and experimental infection, which are not relevant to this work, and they do not attempt to give any indication of the relative importance of humans or other animals in the maintenance of the parasite populations, which is central to our purpose, nor of the abundance of the parasites in humans. Here, we have freely applied our own interpretations to the available evidence.

Only information additional to, or contradicting, the main references is separately referenced. No special effort has been made to seek out primary literature; rather, reviews have been used wherever appropriate. The multi-author works edited by Cox *et al.* (1998) and Palmer *et al.* (1998) have been particularly useful for the addition of numerous details, as has the second edition of Muller's textbook of medical helminthology (Muller 2002).

This second edition, as the first, must be regarded as provisional. The authors would be glad to receive comments, corrections and additions.

Explanation of Terms

Various authors use different spelling for some of the generic names. The publication by ICZN (2001) of those that have been 'listed' facilitates the use of spelling that they have decided, after thorough search of the literature, to be correct. Such names are indicated in the checklist, by the citation, 'ICZN (2001)', where they first appear. Important examples are *Dioctophyme* and *Pthirus*.

Blank lines in entries indicate missing information.

Syn: Major synonyms only are included, especially those which are used in the main reference texts. Vernacular names are sometimes listed.

Status: Indicates some idea of the numbers of reported human cases and their dispersion. A numerical code is tentatively allocated as follows:

1. No more than rare cases anywhere; individual cases are likely to be published
2. Sporadic cases, but nowhere common
3. Common sometimes, but restricted in space and/or time
4. Common in at least large geographical areas
5. Common worldwide, or abundant but geographically restricted
6. Abundant worldwide

Obviously these are semi-quantitative measures, but any attempt to provide greater precision would certainly lead to greater inaccuracy.

Dist: Geographical distribution of human cases by zoogeographical region:

AETH: Aethiopian (or Afrotropical) i.e. Africa south of the Sahara, plus the associated islands

AUS: Australasian (the landmasses east of Wallace's line)

NEA: Nearctic (North America)

NEO: Neotropical (South and Central America)

OR: Oriental (from the Himalayas to Wallace's Line)

PAL: Palaearctic (Africa north of the Sahara, Europe, and Asia north of the Himalayas, with outlying islands including Japan)

Individual countries have been included sometimes, but not systematically. The distribution of the parasite is sometimes indicated where this is much wider than that of known human cases.

Hab: The habitat of parasites in the human host. Where different stages occupy different habitats, this is also indicated.

Hosts: Ideally, non-human hosts which have an essential role in the maintenance of parasite suprapopulations should be listed. In most instances the true maintenance hosts are uncertain, and recorded hosts are listed. The hosts are numbered according to their sequence in the life history, starting with the first intermediate host and ending with the definitive host. Paratenic hosts, in which a parasite survives to be transmitted but does not develop, are indicated with a **P**. We have not attempted to give comprehensive lists of host names, and have used vernacular or technical names indiscriminately, usually following the relevant source.

Trans: Mechanism of infection of humans.

Risk: Any specific risk factor, notably immunodepression, determining human susceptibility to infection.

Zoo stat: Host-specificity status: ideally, the relative role of man (or any other host) in the maintenance of parasite populations can be classified according to one of the six 'ideal' categories in row one of Table 1. The relevant information is lacking for many, if not most, species, so a compromise classification has been used (last row of Table 1) in which 'ideal' categories 1 and 2 have been modified: 3 has been combined with 4, and 5 with 6 as follows:

0: Unknown as a human parasite. (Those forms mentioned in major references are listed following the tabulated summary.)

1: Adventitious or incidental infections in which the transmission stage does not develop or, if it does, is not in a position to be passed on: transmission from man is impossible; suprapopulations are not dependent on humans; usually, but not always, very rare as human parasites; *H. sapiens* is an ecological sink. Parasites such as *Sarcocystis* spp., *Echinococcus* spp., *Trichinella* spp. or pentastomid larvae, which, though viable and transmissible in principle, would depend on humans being eaten in order to be transmitted, have been included in this category.

2: Adventitious or incidental infections in which transmission forms may be produced and released: suprapopulations may be temporarily, but not indefinitely, dependent on humans; *H. sapiens* is at most an ephemeral source of infection, so is an ecological sink. This category probably includes parasites which could theoretically be transmitted from man but are not, so includes some members of the 'ideal' category 1.

3: (Combines 'ideal' categories 3 and 4): shared infections: maintenance hosts are *H. sapiens* and other animals; vertebrate hosts in ecological source system include *H. sapiens*.

4: (Combines 'ideal' categories 5 and 6): core infections: entirely dependent on *H. sapiens*; vertebrate host in ecological source system is *H. sapiens*; any other vertebrate hosts are incidental hosts and ecological sinks.

Even with this simplified classification, numerous decisions have been made arbitrarily, and are subject to review.

Please note that the term 'man' is used in this document to indicate the species *Homo sapiens*, without gender implication.

TABLE 1

Zoo stat: *Classification of parasites according to the role of Man in the maintenance of suprapopulations*

	Infection in Man unknown	*ZOONOSES*				*NON-ZOONOSES*	
		Man has no role	*Man's role irrelevant long term*	*Both Man and other hosts required*	*Man and other hosts maintain populations independently*	*Other hosts' role is irrelevant*	*Other hosts have no role*
Ideal host specificity scale	0	1	2	3	4	5	6
Sink / source	-	Sink	Sink	Sink	Source	Source	Source
Host grade	-	Incidental	Incidental	Shared	Shared	Core	Core
R^* (man – man)	-	0	>0, <1	>0, <1	≥1	≥1	≥1
R^* (other hosts)	≥1	≥1	≥1	>0, <1	≥1	>0, <1	0
Practical host-specificity scale (=**Zoo stat** in checklist)	**0**	**1**	**2**	**3**		**4**	

*: R is the effective long-term reproductive rate (of infection): the number of secondary infections that occur as a consequence of each primary infection

Part 1: Protozoa

CHROMISTA: Bigyra

Blastocystis hominis

Status: 6. Widespread and abundant in most populations
Dist: Cosmopolitan
Hab: Lumen of intestine
Hosts: Identical forms in many mammals; also in birds, reptiles and invertebrates
Trans: (Faeco–oral contamination)
Zoo stat: 3
Boreham and Stenzel 1998

FUNGI?: Microsporidia

Brachiola algerae

Syn: *Nosema algerae*
Status: 1. A single confirmed case
Dist: NEA—Mexico
Hab: Cornea
Hosts: Normally a parasite of mosquitoes
Trans:
Risk: Patient was immunocompetent
Zoo stat: 1
Visvesvara *et al.* 1999; Lowman *et al.* 2000

Brachiola connori

Syn: *Nosema connori*
Status: 1. A single confirmed case
Dist: NEA—USA
Hab: Widely disseminated
Hosts: Known from man only
Trans:
Risk: Immunosuppressed host
Zoo stat: 1
Beaver *et al.* 1984; Deluol and Cenac 1994; Canning 1998; Curry 1999

Brachiola vesicularum

Status: 1. A single case
Dist:
Hab: Muscle
Hosts:
Trans:
Risk:
Zoo stat: 1
Curry 1999

Encephalitozoon cuniculi

Status: 1. Twice in children and a few times in AIDS patients, but up to 8% seropositivity for *Encephalitozoon* sp.
Dist: PAL; AETH; NEA; (Cosmopolitan)
Hab: Widely disseminated
Hosts: A wide range of mammals, especially rabbit and dog; rabbit may be main source of human infection
Trans: Transplacental transmission occurs in carnivore hosts, causing fading-puppy syndrome
Risk: Immunosuppressed host
Zoo stat: 1
Hollister *et al.* 1989, 1996; Deluol and Cenac 1994; Canning 1998; Canning 2001

Encephalitozoon hellem

Status: 1. 'Rarely diagnosed'; only in AIDS patients
Dist: (Cosmopolitan)
Hab: Ocular; extensively in viscera, including kidneys
Hosts: Man, budgerigar *Melopsittacus undulatus*
Trans: May enter through respiratory system
Risk: Immunosuppressed host
Zoo stat: 2
Deluol and Cenac 1994; Hollister *et al.* 1996; Canning 1998; Didier *et al.* 1998; Canning 2001

Encephalitozoon intestinalis

Syn: *Septata intestinalis*
Status: 1. Extremely rare, only in AIDS patients, but may be confused with *E. bieneusi*; much higher prevalence suspected
Dist: (Cosmopolitan)
Hab: Primarily in the intestine, but spreads widely to other organs
Hosts: Dog, pig, cow, goat, donkey
Trans:
Risk: Immunosuppressed host
Zoo stat: 2
Deluol and Cenac 1994; Hartsheerl *et al.* 1995; Canning 1998; Curry 1999

Enterocytozoon bieneusi

Status: 2. Rare: several hundred cases by 1994, but rapidly 'emerging'. Reaches 30% in AIDS patients.
Dist: Cosmopolitan
Hab: Primarily in the intestine, but spreads widely to other organs
Hosts: Man, pig, non-human primates: *Macaca* spp., dog, cat
Trans:
Risk: Immunosuppressed host
Zoo stat: 3
Deluol and Cenac 1994; Canning 1998; Didier *et al.* 1998, Curry 1999

Canning (2001) suggests that pigs and monkeys cannot be sources of human infection, and that this is a 'shared' species.

'Microsporidium' *africanum*

Syn: *Nosema africana*
Status: 1. A single case
Dist: AETH—Botswana
Hab: Cornea
Hosts: Known from man only
Trans:
Zoo stat: 1
Deluol and Cenac 1994; Canning 1998; Curry 1998

'Microsporidium' *ceylonensis*

Syn: *Nosema ceylonensis*
Status: 1. A single case
Dist: OR—Sri Lanka
Hab: Cornea
Hosts: Known from man only
Trans:
Zoo stat: 1
Deluol and Cenac 1994; Canning 1998; Curry 1998

Nosema ocularum

Status: 1. A single case described
Dist: NEA—USA
Hab: Cornea
Hosts: Known from man only
Trans:
Risk: Case was an athymic infant
Zoo stat: 1
Desportes-Livage 1996; Canning 1998

Pleistophora sp.

Most, but not all '*Pleistophora*' records may refer to *Trachipleistophora* spp.
Status: 1.
Dist:
Hab: Muscles
Hosts: Known from man only
Trans:
Risk: Immunosuppressed host
Zoo stat: 1
Canning 1998

Thelohania? sp.

Status: 1. Two cases to 1998
Dist:
Hab: Cardiac muscle, liver, brain, kidney
Hosts: Known from man only
Trans:
Risk: Both cases had HIV infection
Zoo stat: 1
Didier *et al.* 1998

Trachipleistophora anthropophthera

Status: 1. Two patients to date
Dist: NEA—USA
Hab: Brain; widely disseminated elsewhere
Hosts: Known from man only
Risk: Immunosuppressed host
Trans:
Zoo stat: 2
Curry 1999

Trachipleistophora hominis

Most, but not all '*Pleistophora*' infections may refer to this species
Status: 1. A single known case
Dist: AUS—Australia
Hab: Muscles, cornea
Hosts: Known from man only
Risk: Immunosuppressed host
Trans: Carnivory?
Zoo stat: 1
Hollister *et al.* 1996; Canning 1998

Vittaforma corneae

Syn: *Nosema corneum*
Status: 1. Two recorded cases
Dist: NEA—USA
Hab: Cornea; also widely dispersed in an AIDS patient
Hosts: Known from man only
Trans:
Zoo stat: 1
Deluol and Cenac 1994; Canning 1998

METAMONADA: Intestinal flagellates

Chilomastix mesnili

Status: 4. May reach 10%
Dist: Cosmopolitan
Hab: Caecum and colon
Hosts: Identical forms in monkeys and pig
Trans: Faeco–oral contamination
Zoo stat: 3, possibly 4, depending on taxonomy
Beaver *et al.* 1984; Levine 1985

Enteromonas hominis

Status: 2. Nowhere common
Dist: (Cosmopolitan)
Hab: Intestine, especially caecum
Hosts: Identical forms in primates, hamster, rat
Trans: Faeco–oral contamination
Zoo stat: 3 or 4, depending on taxonomy
Beaver *et al.* 1984; Levine 1985

Giardia intestinalis

Syn: *Giardia lamblia*
Giardia duodenalis
Lamblia intestinalis
Status: 6. Endemic in most populations, mainly in children
Dist: Cosmopolitan
Hab: Small intestine
Hosts: Identical forms in primates, carnivores, ungulates, rodents
Trans: Faeco–oral contamination
Zoo stat: 3
Beaver *et al.* 1984

The human strains belong to two related 'assemblages' of genotypes, both shared with several other hosts (Thompson *et al.* 2000).

Retortamonas intestinalis

Status: 2. Rare
Dist: PAL; OR; NEA; NEO; (Cosmopolitan)
Hab: Caecum
Hosts: Described from man only, but may equal species from monkeys, sheep, rabbits, guinea pigs (or insects)
Trans:
Zoo stat: 2
Beaver *et al.* 1984

PARABASALIA

Dientamoeba fragilis

Status: 5. Probably endemic everywhere but prevalence rarely >5%; frequently overlooked
Dist: (Cosmopolitan)
Hab: Lumen of caecum and upper colon
Hosts: 2. Possibly primates
P1? May be paratenic in *Enterobius* eggs
Trans: Unproven, but may be transmitted in egg of *Enterobius vermicularis*; direct contamination unsuccessful
Zoo stat: 4 (Apparently the only genus of parasites restricted to *H. sapiens*)
Beaver *et al.* 1984

Pentatrichomonas hominis

Syn: *Trichomonas hominis*
Status: 5. Endemic in most populations but prevalence rarely >10%
Dist: Cosmopolitan
Hab: Caecal area
Hosts: Identical forms in several other primates, rodents, ox
Trans: (Faeco–oral contamination)
Zoo stat: 3, possibly 4, depending on taxonomy
Beaver *et al.* 1984; Levine 1985

Trichomonas tenax

Status: 6. Probably endemic in all communities; prevalence frequently >20%; 21.4% prevalence in 1250 French subjects.
Dist: Cosmopolitan
Hab: Buccal cavity
Hosts: Identical forms in primates; may be the same as *T. equibuccalis* of equids
Trans: Oral contagion
Zoo stat: 3, possibly 4, depending on taxonomy
Lapierre and Rousset 1973; Beaver *et al.* 1984; Levine 1985; ICZN 2001

Trichomonas vaginalis

Status: 6. Probably endemic in all communities; prevalence frequently >10%
Dist: Cosmopolitan
Hab: Vagina and prostate
Hosts: Only known from man
Trans: Venereal
Zoo stat: 4
Beaver *et al.* 1984

PERCOLOZOA

Naegleria fowleri

Status: 1. Extremely rare
Dist: Cosmopolitan—75% of cases from Australia, 'Czechoslovakia' and USA
Hab: Trophozoites only, in brain
Hosts: Free-living, facultative parasite
Trans: Inhalation of free-living forms in water
Zoo stat: 1
Beaver *et al.* 1984

EUGLENOZOA: Haemoflagellates

Leishmania (Leishmania) aethiopica

Status: 3. Local and rare
Dist: AETH—Ethiopia, Kenya
Hab: Dermis
Hosts: 2. Hyracoidea: *Procavia, Heterohyrax* spp.
1. *Phlebotomus* spp.
Trans: Inoculated by biting sandfly
Zoo stat: 2
Beaver *et al.* 1984; Ashford 1998

Leishmania (Leishmania) amazonensis

Status: 2. Sporadic and rare
Dist: NEO
Hab: Monocytes of dermis
Hosts: 2. Rodents
1. *Lutzomyia* spp.
Trans: Inoculated by biting sandfly
Zoo stat: 2
Beaver *et al.* 1984; Dedet 1993

Leishmania (Leishmania) donovani

Status: 4. Locally abundant and endemic or epidemic
Dist: AETH; OR
Hab: Mononuclear phagocytes, most tissues
Hosts: 2. Rodents: *Arvicanthis* etc. in Africa
1. *Phlebotomus* spp.
Trans: Inoculated by biting sandfly
Zoo stat: 3
Ashford 1998

Leishmania (Leishmania) infantum

Syn: *Leishmania chagasi*
Status: 3. Widespread and endemic but never abundant
Dist: PAL; AETH; NEO
Hab: Mononuclear phagocytes, most tissues
Hosts: 2. Carnivora: *Canis, Vulpes, Cerdocyon* spp. etc.
1. *Phlebotomus, Lutzomyia* spp.
Trans: Inoculated by biting sandfly
Risk: Infancy and malnutrition; immunosuppression
Zoo stat: 1
Ashford 1998

Leishmania (Leishmania) major

Status: 4. Widespread but focal, frequently epidemic
Dist: PAL; AETH
Hab: Dermis
Hosts: 2. Rodents: *Rhombomys, Psammomys, Meriones* etc. N.B. various hosts cited in old Soviet literature predate distinction of *L. turanica*, and are unreliable
1. *Phlebotomus* spp.
Trans: Inoculated by biting sandfly
Zoo stat: 2
Beaver *et al.* 1984; Ashford 1998

Mostly zoonotic, but there is evidence of ephemeral anthropogenic spread.

Leishmania (*Leishmania*) *mexicana*

Status: 2. Sporadic cases widespread but focal
Dist: NEO
Hab: Monocytes of dermis, superficial cartilage
Hosts: 2. Rodents
1. *Lutzomyia* spp.
Trans: Inoculated by biting sandfly
Zoo stat: 2
Beaver *et al.* 1984; Dedet 1993

Leishmania (*Leishmania*) *tropica*

Includes *Leishmania killicki*
Status: 4. Locally abundant and endemic or epidemic
Dist: PAL; OR; AETH
Hab: Dermis
Hosts: 2. Dog (rare); Hyracoidea: *Procavia* (AETH)
1. *Phlebotomus* spp.
Trans: Inoculated by biting sandfly
Zoo stat: 3
Beaver *et al.* 1984; Sang *et al.* 1992; Ashford 1998

Following the discovery of parasites close to *L. tropica* in zoonotic foci in Africa, this species is removed from the group of **non-**zoonoses **not** found in Africa. The African forms, sometimes grouped as *L. killicki*, provide a potential ancestral source, especially as they are grouped at the base of the *L. tropica* cladogram. These zoonotic (proven or strongly suspected) *L. tropica* strains have been found in Tunisia, where the reservoir host is suspected to be the gundi *Ctenodactylus gundi*, and in Kenya and Namibia, where the probable reservoir hosts are *Procavia* spp. Anthroponotic transmission has not been suspected in Africa.

Strains of *L. tropica* from Jordan and Israel are also suspected to be transmitted zoonotically. According to epidemiological models, anthroponotic endemic maintenance of *L. tropica* requires large, dense, human populations and massive numbers of anthropophilic sandflies, so it must be of relatively recent origin (Ashford 1997).

Leishmania (*Viannia*) *braziliensis*

Status: 3. Generally sporadic; occasional epidemics
Dist: NEO
Hab: Monocytes, dermis and superficial mucosae
Hosts: 2. Various, but main hosts unknown
1. *Lutzomyia* spp.
Trans: Inoculated by biting sandfly
Zoo stat: 2
Beaver *et al.* 1984; Dedet 1993

Leishmania (*Viannia*) *guyanensis*

Status: 3. High incidence focally
Dist: NEO
Hab: Monocytes, dermis
Hosts: 2. Edentata: *Choloepus*, *Bradypus*
1. *Lutzomyia* spp.
Trans: Inoculated by biting sandfly
Zoo stat: 2
Beaver *et al.* 1984; Dedet 1993

Leishmania (*Viannia*) *lainsoni*

Status: 1. Some 20 cases between 1981 and 1990
Dist: NEO
Hab: Monocytes, dermis
Hosts: 2. Rodents: *Agouti paca*
1. *Lutzomyia* spp.
Trans: Inoculated by biting sandfly
Zoo stat: 2
Dedet 1993

Leishmania (Viannia) naiffi

Status: 1. Very few human cases
Dist: NEO
Hab: Monocytes, dermis
Hosts: 2. Edentata: *Dasypus*
1. *Lutzomyia* spp.
Trans: Inoculated by biting sandfly
Zoo stat: 2
Dedet 1993

Leishmania (Viannia) panamensis

Status: 3. High incidence focally
Dist: NEO
Hab: Monocytes, dermis
Hosts: 2. Edentata: *Choloepus*; various minor hosts
1. *Lutzomyia* spp.
Trans: Inoculated by biting sandfly
Zoo stat: 2
Beaver *et al.* 1984; Dedet 1993

Leishmania (Viannia) peruviana

Status: 3. High incidence, but very focal
Dist: NEO
Hab: Monocytes, dermis
Hosts: 2. Domestic dog; presumably others, unknown
1. *Lutzomyia* spp.
Trans: Inoculated by biting sandfly
Zoo stat: 2
Beaver *et al.* 1984; Dedet 1993

Leishmania (Viannia) shawi

Status: 1. Very rare human cases
Dist: NEO
Hab: Monocytes, dermis
Hosts: 2. Various; main hosts ?Edentata: *Choloepus*, *Bradypus*
1. *Lutzomyia* spp.
Trans: Inoculated by biting sandfly
Zoo stat: 2
Dedet 1993

Leishmania sp.

Status: 1. Two (five?) cases reported
Dist: NEO—Martinique
Hab: Dermis
Hosts: Insects?
Trans: Vector-borne?
Risk: Immunosuppressed host?
Zoo stat: 2
Dedet *et al.* 1995; Noyes *et al.* 2002

Trypanosoma brucei gambiense

Status: 3. Locally common, with epidemics
Dist: AETH
Hab: Blood
Hosts: 2. Man appears to be the main host; other animals are incidental hosts
1. *Glossina* spp.
Trans: Inoculated by biting tsetse fly
Zoo stat: 4
Beaver *et al.* 1984; ICZN 2001

Trypanosoma brucei rhodesiense

Status: 3. Usually sporadic, occasional epidemics
Dist: AETH
Hab: Blood
Hosts: 2. Ungulates
1. *Glossina* spp.
Trans: Inoculated by biting tsetse fly
Zoo stat: 2
Beaver *et al.* 1984

According to Truc *et al.* (1997), there seems to be one group of *T. brucei* strains, clearly identifiable as '*T. b. gambiense* Group 1', which occurs predominantly in humans in West Africa, but occasionally in wild and domestic animals. Other stocks, isolated mainly from animals, cannot be distinguished as *T. b rhodesiense* or *T. b. brucei.* The only defining feature of '*T. b. rhodesiense*' is infectivity to man, and stocks with this feature seem to be polyphyletic within *T. b. brucei*. The findings of Komba *et al.* (1998) are compatible with this conclusion.

Trypanosoma cruzi

Status: 5. Thirty-five million people 'exposed to infection'
Dist: NEO—17 countries; NEA
Hab: Amastigotes in myocytes, trypomastigotes in blood
Hosts: 2. Multiple mammalian orders
1. Triatomidae: *Triatoma*, *Rhodnius* spp.
Trans: Trypomastigote from triatomid faeces penetrates feeding wound or conjunctiva
Zoo stat: 3
Beaver *et al.* 1984

Trypanosoma rangeli

Status: 2. 'Demonstrated frequently in man; several hundred cases reported'
Dist: NEO—Eight countries, possibly 14
Hab: Blood
Hosts: 2. Multiple mammalian orders
1. *Rhodnius* spp.
Trans: Trypomastigote injected by feeding triatomid
Zoo stat: 2
Beaver *et al.* 1984

Trypanosoma sp. (*cf. T. lewisi*)

Probably two species
Status: 1. At least five cases to 1977
Dist: OR—Peninsular Malaysia, India
Hab: Blood
Hosts: 2. Unknown; *Trypanosoma* spp. in South-east Asian primates are apparently different
1. Unknown
Trans: (Vector-borne)
Zoo stat: 2
Weinman 1977

AMOEBOZOA

Acanthamoeba castellanii

Syn: *Acanthamoeba rhysoides*
Status: 2. The most frequent *Acanthamoeba* identified in both granulomatous amoebic encephalitis and amoebic keratitis
Dist: (Cosmopolitan)
Hab: Trophozoites and cysts in brain or cornea
Hosts: Free-living, facultative parasite
Trans: Direct entry of trophozoites from water
Risk: Wearing of contact lenses; HIV infection increases risk of encephalitis
Zoo stat: 1
Warhurst 1985; John 1998

Acanthamoeba species may act as reservoir hosts for *Legionella pneumophila* causing Pontiac fever or humidifier fever. Seventeen species of *Acanthamoeba* are known, of which seven have been associated with human infection (John 1998).

Acanthamoeba culbertsoni

Status: 1. Extremely rare
Dist: (Cosmopolitan)
Hab: Trophozoites and cysts in brain and cornea
Hosts: Free-living, facultative parasite
Trans: Inhalation of free-living forms in water; also associated with contact lenses
Zoo stat: 1
Beaver *et al.* 1984

Acanthamoeba griffini

Status: 1. A single case, 1996
Dist: NEA—USA; (parasite presumed Cosmopolitan)
Hab: Cornea
Hosts: Free-living, facultative parasite
Trans: Associated with contact lenses
Zoo stat: 1
Ledee *et al.* 1996

Acanthamoeba hatchetti

Status: 1. Extremely rare
Dist: (Cosmopolitan)
Hab: The only corneal *Acanthamoeba* not to have been cultured from clinical material
Hosts: Free-living, facultative parasite
Trans: Direct entry of trophozoites from water; associated with contact lenses
Zoo stat: 1
John 1998

Acanthamoeba palestinensis

Status: 1. A single case mentioned
Dist: (Cosmopolitan)
Hab: Trophozoites and cysts in brain tumour
Hosts: Free-living, facultative parasite
Trans:
Zoo stat: 1
John 1998

Acanthamoeba polyphaga

Status: 1. Extremely rare
Dist: (Cosmopolitan)
Hab: Trophozoites and cysts in cornea
Hosts: Free-living, facultative parasite
Trans: Direct entry of trophozoites from water; associated with contact lenses
Zoo stat: 1
John 1998

Balamuthia mandrillaris

Status: 1. Initially, 16 cases were identified retrospectively; 'infections continue to be reported'
Dist: (Cosmopolitan)
Hab: Trophozoites and cysts in tissues; causes granulomatous amoebic encephalitis
Hosts: Free-living, but rarely isolated in nature
Trans:
Zoo stat: 1
John 1998

Endolimax nana

Status: 6. Abundant in most communities
Dist: Cosmopolitan
Hab: Lumen of caecum and colon
Hosts: Identical forms in primates, rodents
Trans: Cysts ingested by faeco–oral contamination
Zoo stat: 3
Beaver *et al.* 1984

Entamoeba chattoni

Includes *Entamoeba polecki*
Status: 1. Extremely rare, more common locally in Papua New Guinea
Dist: (Cosmopolitan)
Hab: Lumen of caecum and colon
Hosts: The commonest amoeba of primates, also abundant in pigs
Trans: Contamination with primate or pig faeces
Zoo stat: 2
Sargeaunt *et al.* 1992; ICZN 2001

Sargeaunt *et al.* consider *E. polecki* to be a *nomen nudum*, and that all such cases should be referred to *E. chattoni.*

Entamoeba coli

Status: 6. Abundant in most communities
Dist: Cosmopolitan
Hab: Lumen of caecum and colon
Hosts: Primates, pig
Trans: Cysts ingested by faeco–oral contamination
Zoo stat: 3
Beaver *et al.* 1984

Entamoeba dispar

Status: 6. Common in many communities
Dist: Cosmopolitan
Hab: Lumen of large intestine; rarely invasive
Hosts: Man is the only maintenance host
Trans: Cysts ingested by faeco–oral contamination
Zoo stat: 4
Diamond and Clark 1993

Of the two morphologically similar species of *Entamoeba* with four-nucleate cysts in man, pathogenic strains are predominantly *E. histolytica*, whereas *E. dispar* is rarely if ever invasive. About nine in 10 infections are said to be of *E. dispar*. The two species can only be distinguished biochemically; *E.dispar* rarely if ever produces an immune response, whereas *E. histolytica* usually does.

Entamoeba gingivalis

Status: 6. Abundant but rarely looked for; 36.2% prevalence in 1250 French subjects
Dist: (Cosmopolitan); few if any records from Asia or Africa
Hab: Surfaces of buccal mucosa and teeth
Hosts: Identical forms in other primates
Trans: Direct or indirect oral contact
Zoo stat: 3, or 4, depending on taxonomy
Lapierre and Rousset 1973; Beaver *et al.* 1984

Entamoeba hartmanni

Status: 6. Common in most communities, but rarely identified
Dist: Cosmopolitan
Hab: Lumen of caecum and colon
Hosts: Primates, others unknown
Trans: Cysts ingested by faeco–oral contamination
Zoo stat: 3
Beaver *et al.* 1984

Entamoeba histolytica

Status: 5. Common in many communities
Dist: Cosmopolitan
Hab: Lumen and mucosa of large intestine; invades liver, lungs etc.
Hosts: Man is the only maintenance host
Trans: Cysts ingested by faeco–oral contamination
Zoo stat: 4
Beaver *et al.* 1984

Since this species is usually pathogenic, producing trophozoites or immature cysts in faeces, it is hard to see how it can be maintained in human populations without the intervention of an alternative host.

Entamoeba moshkovskii

Syn: *Entamoeba histolytica* Laredo strain
Status: 1. Infrequently isolated from humans
Dist: AETH—S. Africa; NEA—USA; OR—Bangladesh; PAL—Italy;
Hab: (Lumen of large intestine)
Hosts: Normally free-living; one of at least six strains can infect humans
Trans: (Cysts ingested by faeco–oral contamination)
Zoo stat: 2
Martinez-Palomo and Cantellano 1998; Haque *et al.* 1998

Iodamoeba buetschlii

Status: 6. Frequently abundant
Dist: Cosmopolitan
Hab: Lumen of caecum and colon
Hosts: Identical forms in primates, pig
Trans: Cysts ingested by faeco–oral contamination
Zoo stat: 3
Beaver *et al.* 1984

DINOZOA: Dinoflagellates

Pfiesteria piscicida

Status: 1. A new 'parasite'; dozens of cases, but very localised
Dist: NEA—East coast of USA
Hab: Skin lesions
Hosts: Normally free living, or fish parasite
Trans: Exposure to free living forms in estuarine waters
Zoo stat: 2
Samet *et al.* 2001

The parasitic nature of this organism is doubtful; it has been called an 'ambush-predator'; secreted toxins cause skin lesions, and organisms feed on necrotic tissue.

SPOROZOA

Babesia divergens

Cox argues cogently that human cases are of this species rather than *Babesia bovis*
Status: 1. Twenty-one cases recorded
Dist: PAL
Hab: Erythrocytes
Hosts: 2. Ungulates
1. Acari: Ixodidae
Trans: Injected in saliva of feeding tick
Risk: Splenectomy
Zoo stat: 2
Beaver *et al.* 1984; Cox 1998

Babesia gibsoni

Status: 1. Five cases in northern California to 1998; 'A *B. gibsoni*-like parasite is known to infect humans...'
Dist: NEA—'...along the Pacific coast of the USA'
Hab: Erythrocytes
Hosts: 2. Dog
1. Acari: Ixodidae
Trans: Injected with saliva of feeding tick
Zoo stat: 2
Telford and Spielman 1998

Babesia microti

Status: 2. Hundreds of cases to 1998; three cases in Europe
Dist: NEA—North-eastern and upper mid-western USA; PAL—Europe, Japan, Taiwan
Hab: Erythrocytes
Hosts: 2. Rodents: *Microtus* spp.
1. Acari: Ixodidae
Trans: Injected with saliva of feeding tick
Zoo stat: 2
Beaver *et al.* 1984; Cox 1998; Telford and Spielman 1998

The Japanese parasites differ from the American ones, and are named Kobe and Hobetsu strains (Tsuji *et al.* 2001).

Cryptosporidium baileyi

Status: 1. First case reported 1991
Dist: PAL—Czech Republic
Hab: Throughout the alimentary and respiratory systems and gall and urinary bladders
Hosts: Domestic chicken
Trans: Ingestion of oocysts
Risk: The case was immunosuppressed following kidney transplant
Zoo stat: 2
Ditrich *et al.* 1991

Cryptosporidium canis

Originally reported as the canine genotype of *Cryptosporidium parvum*

Status: 1. First case reported as *C. canis*, 2001
Dist: NEA—USA; NEO—Peru; PAL—Switzerland
Hab: Intestine
Hosts: Only known from dog and human
Trans: Ingestion of oocysts
Risk: The case was HIV-positive
Zoo stat: 2
Fayer *et al.* 2001

Cryptosporidium felis

Status: 1. Fourteen cases to 2002
Dist: AETH; NEA—USA; NEO; PAL—Italy, Switzerland
Hab: Intestine
Hosts: Cat; once from a cow
Trans: Ingestion of oocysts
Risk: Increased incidence of serious disease in AIDS
Zoo stat: 2
Caccio *et al.* 2002

Cryptosporidium meleagridis

Status: 1. Twenty-one cases recognised to 2002
Dist: AETH—Kenya; PAL—Japan, Switzerland (Parasite cosmopolitan)
Hab: May occur throughout the intestine, in surface membrane of enterocytes
Hosts: Birds
Trans: Ingestion of oocysts
Risk: May infect humans regardless of immunological status
Zoo stat: 2
Yagita *et al.* 2001

Cryptosporidium muris

Status: 1.
Dist: AETH—Kenya; PAL—France (Parasite cosmopolitan)
Hab: In mice, mainly in the stomach
Hosts: Rodents
Trans: Ingestion of oocysts
Risk: The patient was suffering from AIDS
Zoo stat: 2
Gatei *et al.* 2002

Cryptosporidium parvum

Status: 5. 'Eight cases by 1982'. Typical point prevalence about 0.1%, but lifetime prevalence probably 100% in many communities
Dist: Cosmopolitan
Hab: May occur throughout the intestine, in surface membrane of enterocytes
Hosts: Many mammals may maintain populations
Trans: Ingestion of oocysts
Risk: Increased incidence of serious disease in AIDS
Zoo stat: 3
Beaver *et al.* 1984; Awad El Kariem *et al.* 1995

Strains from cattle and sheep are mostly different from strains of human origin, suggesting that most transmission is interhuman. The strain maintained in humans is likely to be described as a separate species (Thompson and Chalmers (2002).

Cyclospora cayetanensis

Status: 3. Poorly known; sometimes common
Dist: Cosmopolitan
Hab: Enterocytes of small intestine
Hosts: Known from man, baboon and chimpanzee
Trans: Ingestion of sporulated oocysts with contaminated food or water
Zoo stat: 3, pending further information
Ashford 1979; Smith *et al.* 1996

Isospora belli

Status: 2. Very rare; outbreaks are exceptional
Dist: (Cosmopolitan)
Hab: Upper small intestine; in lymph nodes in immunosuppressed patients
Hosts: Man is the only known host
Trans: Ingestion of sporulated oocysts
Risk: Increased incidence in AIDS
Zoo stat: 4
Beaver *et al.* 1984

This must be the rarest parasite of man that (apparently) has no other host. The persistence of suprapopulations of *I. belli* is unexplained.

Plasmodium falciparum

Status: 5. Lifetime prevalence 100% in many countries
Dist: Cosmopolitan
Hab: Schizonts in liver cells; other stages in erythrocytes
Hosts: 2. Man only
1. Diptera: Culicidae: *Anopheles* spp.
Trans: Sporozoite injected by feeding mosquito
Zoo stat: 4
Beaver *et al.* 1984; ICZN 2001

P. falciparum is represented by *Plasmodium reichenowi* in *Pan* and *Gorilla*, but has no other primate, or even mammalian, homologue. Ayala *et al.* (1999) suggest that the primates and parasites speciated simultaneously around 10 million years ago.

Plasmodium knowlesi

Status: 1. A single confirmed natural human infection, but was once used for malaria therapy
Dist: OR—Malaysia
Hab: Only known in man from blood stages
Hosts: 2.
1. Diptera: Culicidae: *Anopheles hackeri*
Trans: Sporozoite injected by feeding mosquito
Zoo stat: 2
Coatney *et al.* 1971

Plasmodium malariae

Syn: *Plasmodium rodhaini* of *Pan* is probably the same species and *Plasmodium brasilianum* of Neotropical primates is possibly a recently derived race
Status: 4. Reaches 10% prevalence in some communities
Dist: Cosmopolitan
Hab: Schizonts in liver cells; other stages in erythrocytes
Hosts: 2. Man, *Pan*
1. Diptera: Culicidae: *Anopheles* spp.
Trans: Sporozoite injected by feeding mosquito
Zoo stat: 3
Beaver *et al.* 1984

According to Peters *et al.* (1976), *P. malariae* is found in both *H. sapiens* and *Pan* spp. (*P. rodhaini*) as well as in New World monkeys (*P. brasilianum*), in which it was probably introduced as a 'zoonosis in reverse'. Incongruously, the closest relative is said to be *Plasmodium inui* of Asian monkeys: there is no known vicariant in Asian apes. The identity of *P. malariae* and *P. brasilianum* is confirmed genetically by Escalante *et al.* (1995), who suppose that humans were only exposed when they reached the New World. This strange notion was questioned by Ayala *et al.* (1999), who imply that *P. malariae* was acquired by a pre-hominid ancestor, from a non-primate host.

Plasmodium ovale

Syn: Probably synonymous with *Plasmodium schwetzi* of *Pan troglodytes* and *Gorilla gorilla* in West, but not East Africa
Status: 4. Prevalence rarely >1%
Dist: AETH—'Mostly in tropical Africa, principally on the west coast'
Hab: Hypnozoites and schizonts in liver cells; other stages in erythrocytes
Hosts: 2. Man only
1. Diptera: Culicidae: *Anopheles* spp.
Trans: Sporozoite injected by feeding mosquito
Zoo stat: 3, or 4, depending on taxonomy
Coatney *et al.* 1971; Beaver *et al.* 1984

See under *P. vivax* for the views of Peters *et al.* (1976).

Ayala *et al.* (1999) imply that *P. ovale* was acquired by a pre-hominid ancestor, from a non-primate host.

Plasmodium vivax

Syn: Thought to be synonymous with *Plasmodium simium* of Neotropical primates: another 'zoonosis in reverse'
Status: 5. The commonest malarial parasite of man in non-tropical areas; lifetime prevalence frequently 100%
Dist: Cosmopolitan, but absent from West Africa
Hab: Hypnozoites and schizonts in liver cells; other stages in erythrocytes
Hosts: 2. Man only
1. Diptera: Culicidae: *Anopheles* spp.
Trans: Sporozoite injected by feeding mosquito
Zoo stat: 4
Beaver *et al.* 1984

For Peters *et al.* (1976), *P. vivax* is the most ancient human malarial parasite, dating to the early primates, and having common ancestry with *P. ovale* of *Homo, P. schwetzi* of *Pan* and *Gorilla*, *Plasmodium silvaticum* of *Pongo*, *Plasmodium eylesi* of *Hylobates*, and various other *Plasmodium* species of cercopithecines, notably *Plasmodium cynomolgi*.

Escalante *et al.* (1995) suppose that humans were only exposed when they reached the New World. This strange notion was questioned by Ayala *et al.* (1999), who imply that *P. vivax* was acquired by a pre-hominid ancestor, from a non-primate host.

Qari *et al.* (1993) found *P. vivax*-like parasites in Papua New Guinea and Brazil, in which the circumsporozoite protein gene sequence differed from that of normal *P. vivax*, but was identical with that of *Plasmodium simiovale,* which was first described from *Macaca sinica* in Sri Lanka. The identity of the '*P. vivax*-like' parasites and *P. simiovale* is supported by Ayala *et al.* (1999).

Occasional infections with *P. ovale* were previously described from Papua New Guinea; conceivably these were the *simiovale*-like parasite.

Sarcocystis hominis

Syn: *Sarcocystis fusiformis*
Sarcocystis bovihominis
Status: 3. Common throughout the world (*sic*)
Dist: Cosmopolitan; more common in Europe than elsewhere
Hab: Enterocytes
Hosts: 2. Man only
1. Ox and presumably zebu
Trans: Ingestion of bradyzoites in infected beef
Zoo stat: 4
Beaver *et al.* 1984; Levine 1985; Dubey *et al.* 1989; ICZN 2001

Sarcocystis suihominis

Syn: *Sarcocystis miescheriana pro parte*
Status: 3. Apparently quite common
Dist: Cosmopolitan; more common in Europe than elsewhere
Hab: Enterocytes
Hosts: 2. Man only
1. Pig
Trans: Ingestion of bradyzoites in infected pork
Zoo stat: 4
Beaver *et al.* 1984; Levine 1985

Sarcocystis 'lindemanni"

Probably at least seven species
Status: 1. At least 60 cases to 1992; 21 of 100 autopsies positive in Malaysia
Dist: Cosmopolitan; each species must have its own distribution
Hab: Sarcocysts in muscle
Hosts: Unknown
Trans: (Ingestion of sporocysts in carnivore faeces)
Zoo stat: 1
Beaver *et al.* 1984; Wong and Pathmanathan 1992

PART 1: PROTOZOA

Toxoplasma gondii

Status: 6. One of the most widespread and abundant of all parasites; lifetime prevalence frequently >50%
Dist: Cosmopolitan
Hab: Intracellular in many cells, particularly macrophages and myocytes
Hosts: 2. Felidae
1. Most mammals and birds
Trans: Ingestion of oocysts by contamination with cat faeces; ingestion of bradyzoites in meat (especially pig or sheep) or brain; transplacental
Zoo stat: 1
Beaver *et al.* 1984

CILIATA

Balantidium coli

Status: 2. Nowhere common, exceptionally epidemic
Dist: (Cosmopolitan)
Hab: Lumen of large intestine; freely invades tissue
Hosts: Pig, primates; occasionally maintained in humans
Trans: Ingestion of cysts by contamination with pig or monkey faeces
Zoo stat: 3
Beaver *et al.* 1984; Zaman 1998; ICZN 2001

Part 2: Trematoda

CLINOSTOMIDAE

Clinostomum complanatum

Status: 1. Fifteen cases in Japan to 1997
Dist: PAL—Japan, Korea, Middle East
Hab: Attached to wall of pharynx
Hosts: 3. Piscivorous birds
2. Many freshwater fish
1. Snails: *Helisoma*, *Lymnaea* spp.
Trans: Ingestion of metacercaria in fish
Zoo stat: 1
Beaver *et al.* 1984; Coombs and Crompton 1991; Yoshida 1997; Muller 2002

CYATHOCOTYLIDAE

Prohemistomum vivax

Status: 1. A single human case, 1941
Dist: PAL—Egypt
Hab: Small intestine
Hosts: 3. Fish-eating birds, dog, cat
2. Brackish-water fish: *Clarias*, *Mugil*, *Tilapia* spp.
1. Snails: *Cleopatra*, *Melanopsis* spp.
Trans: Ingestion of metacercaria in fish
Zoo stat: 2
Muller 2000

DIPLOSTOMIDAE

Alaria alata

Status: 1. Occasional cases in Middle East
Dist: NEA—USA, Canada; PAL—Europe, Middle East
Hab: Mesocercaria paratenic, embedded in tissues
Hosts: 3. Carnivora: dog
P2.
2. Amphibia
1. Snails
Trans: Ingestion of hosts 2 or P2
Zoo stat: 1
Muller 2000, 2002

Alaria americana

Status: 1. Two recorded cases
Dist: NEA—USA, Canada
Hab: Mesocercaria paratenic in numerous organs
Hosts: 3. Adults in Carnivora
P2. Amphibia, snakes
2. Mesocercaria in Amphibia
1. Asexual stages in Mollusca: *Helisoma* spp.
Trans: Ingestion of mesocercaria in frogs
Zoo stat: 1
Beaver *et al.* 1984; Coombs and Crompton 1991

Alaria marcianae

Status: 1. A single human case
Dist: NEA—USA, Canada
Hab: Mesocercaria paratenic, embedded in tissues
Hosts: 3. Carnivora: dog, *Vulpes*
P2. *Procyon*, *Didelphis*
2. Amphibia and snakes
1. Snails: possibly *Planorbis*
Trans: Ingestion of hosts 2 or P2
Zoo stat: 1
Beaver *et al.* 1984; Coombs and Crompton 1991; Muller 2000

Diplostomum spathaceum

Status: 1.
Dist: PAL
Hab: Cercaria penetrates skin and migrates to eye
Hosts: 3. Piscivorous birds: *Larus* spp.; 38 avian hosts
2. Freshwater fish
1. Snail: *Lymnaea* sp.
Trans: Cercaria actively invades skin
Zoo stat: 1
Coombs and Crompton 1991; Smyth 1995

Neodiplostomum seoulense

Syn: *Fibricola seoulensis*
Status: 1. Twenty-five human cases recorded
Dist: PAL—Korea
Hab: Adult in alimentary tract
Hosts: 3. *Rattus* spp.
P2. Snake: *Rhabdophis* sp.
2. Frogs: *Rana* spp.
1. Snails: *Hippeutis, Segmentina* spp.
Trans: Ingestion of metacercaria in flesh of frog or snake
Zoo stat: 2
Coombs and Crompton 1991; Muller 2002

Pharyngostomum flapi

Status: 1.
Dist: PAL—Egypt
Hab: Adult in small intestine
Hosts: 3.
2.
1.
Trans: Ingestion of metacercaria in fish
Zoo stat: 2
Muller 2002

STRIGEIDAE

Cotylurus japonicus

Status: 1.
Dist: PAL—China
Hab:
Hosts: 3.
2.
1.
Trans:
Zoo stat: 2
Muller 2002

BRACHYLAIMIDAE

Brachylaima cribbi

Status: 1. A single record
Dist: AUS—Australia (parasite probably of European origin)
Hab: Eggs in faeces
Hosts: 3. Only known from man
2. Helicid snails
1. Helicid snail: *Theba, Cornuella, Helix* spp.
Trans: Ingestion of raw snail
Zoo stat: 2
Butcher and Grove 2001

GYMNOPHALLIDAE

Gymnophalloides seoi

Status: 2. Locally common in Korea
Dist: PAL—Korea
Hab: Adults in intestine
Hosts: 3. Shorebirds, e.g. Oystercatcher *Haematopus ostralegus*
2. Oyster *Crassostrea gigas*
1.
Trans: Ingestion of metacercaria in raw oyster
Zoo stat: 2
Lee and Chai 2001

ISOPARORCHIIDAE

Isoparorchis hypselobagri

Status: 1.
Dist: PAL; OR; AUS
Hab: (Alimentary tract); adult worm expelled following treatment
Hosts: 3. Adult in swim bladder of freshwater fish
2.
1. Snail: *Posticobia* sp.
Trans: Ingestion of metacercaria in fish
Zoo stat: 1
Coombs and Crompton 1991

SCHISTOSOMATIDAE

Austrobilharzia terrigalensis

Status: 1.
Dist: AUS
Hab: Abortive schistosomule in skin
Hosts: 2. Aquatic birds
1. Snail: *Littorina* sp.
Trans: Cercaria actively invades skin
Zoo stat: 1
Coombs and Crompton 1991

Bilharziella polonica

Status: 1.
Dist: PAL
Hab: Abortive schistosomule in skin
Hosts: 2. Aquatic birds
1. Various pulmonate snails
Trans: Cercaria actively invades skin
Zoo stat: 1
Coombs and Crompton 1991

Gigantobilharzia huttoni

Status: 1.
Dist: NEA
Hab: Abortive schistosomule in skin
Hosts: 2. Aquatic birds: *Pelecanus* spp.
1. Snail: *Haminoea* sp.
Trans: Cercaria actively invades skin
Zoo stat: 1
Coombs and Crompton 1991

Gigantobilharzia sturniae

Status: 2. The commonest cause of schistosomal dermatitis in Japan
Dist: PAL—Japan
Hab: Abortive schistosomule in skin
Hosts: 2. Various passerine birds
1. Snails: *Segmentina* sp., *Polypilis* sp.
Trans: Cercaria actively invades skin
Zoo stat: 1
Yamaguchi 1981; Coombs and Crompton 1991

Heterobilharzia americana

Status: 1.
Dist: NEA—USA
Hab: Abortive schistosomule in skin
Hosts: 2. Many wild and domestic animals
1. Snail: *Pseudosuccinea, Lymnaea* spp.
Trans: Cercaria actively invades skin
Zoo stat: 1
Coombs and Crompton 1991

Microbilharzia variglandis

Status: 1.
Dist: NEA
Hab: Abortive schistosomule in skin
Hosts: 2. Birds
1.
Trans: Cercaria actively invades skin
Zoo stat: 1
Muller 2002

Orientobilharzia turkestanica

Status: 1.
Dist: PAL
Hab: Abortive schistosomule in skin
Hosts: 2. Many wild and domestic animals
1. Snail: *Lymnaea* sp.
Trans: Cercaria actively invades skin
Zoo stat: 1
Coombs and Crompton 1991

Schistosoma bovis

Status: 3. An important cause of cercarial dermatitis in Sardinia
Dist: AETH; PAL
Hab: Abortive schistosomule in skin
Hosts: 2. Cattle, sheep, goats
1. Snails: *Bulinus* spp.
Trans: (Cercaria actively invades skin)
Zoo stat: 1
Biocca 1960; Coombs and Crompton 1991; ICZN 2001

Schistosoma haematobium

Status: 5. Widespread and abundant; estimated 39 million cases, 1980
Dist: AETH; PAL
Hab: Adult in veins of urinary bladder
Hosts: 2. Occasional infections in primates are probably mainly incidental, though apparently endemic in places
1. Snails: *Bulinus* spp.
Trans: Cercaria actively penetrates skin
Zoo stat: 4
Malek 1980a; Coombs and Crompton 1991; Jordan *et al.* 1993

Schistosoma intercalatum

Status: 3. Range very restricted, but locally common
Dist: AETH—Forested areas of West and Central Africa
Hab: Adult in mesenteric blood vessels
Hosts: 2. Man is only natural host known
1. Snail: *Bulinus* sp.
Trans: Cercaria actively penetrates skin
Zoo stat: 4
Coombs and Crompton 1991; Jordan *et al.* 1993

Schistosoma japonicum

Status: 4. 1948: 138 counties in China; estimated 46 million cases, 1980
Dist: OR—Indonesia, Malaysia, Philippines, Thailand; PAL—China, Japan
Hab: Adult in mesenteric blood vessels
Hosts: 2. More than 30 mammal species; reservoir host: cattle (China), pig (Philippines), various (Sulawesi, Indonesia)
1. Snail: *Oncomelania* sp.
Trans: Cercaria actively penetrates skin
Zoo stat: 3
Malek 1980a; Sobhon and Upatham 1990; Coombs and Crompton 1991

Schistosoma malayensis

Status: 1. Rare, abortive infection of tribal people in very restricted range
Dist: OR—Malaysia
Hab: Known only from eggs in liver
Hosts: 2. *Rattus muelleri*
1. Snails: *Robertsiella* spp.
Trans: (Cercaria actively penetrates skin)
Zoo stat: 1
Sobhon and Upatham 1990; Coombs and Crompton 1991

Schistosoma mansoni

Status: 5. Holo-endemic in many communities; estimated 29 million cases in 1980
Dist: AETH; PAL; NEO
Hab: Adult in mesenteric blood vessels
Hosts: 2. Probably maintained by baboons in very localised foci; the roles of other primates and of rodents in Africa have not been fully established. Rodents maintain secondary foci in South America
1. Snails: *Biomphalaria* spp.
Trans: Cercaria actively penetrates skin
Zoo stat: 4, possibly 3
Malek 1980a; Coombs and Crompton 1991; Ouma and Fenwick 1991

Schistosoma mattheei

Status: 3. Limited distribution but locally common, with prevalence reaching 40%
Dist: AETH—Southern Africa
Hab: Known from eggs in rectal biopsy
Hosts: 2. Horses, cattle, sheep, zebra, baboons, antelopes
1. Snails: *Bulinus* spp.
Trans: Cercaria actively penetrates skin
Zoo stat: 1
Coombs and Crompton 1991; Jordan *et al.* 1993

Schistosoma mekongi

Status: 3. Very localised, but prevalence may reach >10%
Dist: OR—Laos, Cambodia, Thailand
Hab: (Mesenteric blood vessels)
Hosts: 2. Dog
1. Snail: *Neotricula* sp.
Trans: Cercaria actively penetrates skin
Zoo stat: 2
Sobhon and Upatham 1990; Coombs and Crompton 1991

Schistosoma rodhaini

Status: 1. A single human case record in 'Zaire'
Dist: AETH—Congo, Burundi, Uganda, Kenya
Hab: (Venous blood vessels)
Hosts: 2. Rodents, Carnivora
1. Snails: *Biomphalaria* spp.
Trans: Cercaria actively penetrates skin
Zoo stat: 2
Beaver *et al.* 1984; Coombs and Crompton 1991; Jordan *et al.* 1993

Schistosoma sinensium

Status: 1. 'A few cases'
Dist: PAL—China
Hab:
Hosts: 2.
1. Snail: *Tricula* sp.
Trans: (Cercaria actively penetrates skin)
Zoo stat: 1
Muller 2002

Schistosoma spindale

Status: 1.
Dist: OR
Hab: Abortive schistosomule in skin
Hosts: 2. Many species of wild and domestic mammals
1. Snail: *Indoplanorbis* sp.
Trans: Cercaria actively invades skin
Zoo stat: 1
Coombs and Crompton 1991

Schistosomatium douthitti

Status: 1.
Dist: NEA
Hab: Abortive schistosomule in skin
Hosts: 2. Rodents
1. Snails: *Lymnaea* spp.
Trans: Cercaria actively invades skin
Zoo stat: 1
Coombs and Crompton 1991

Trichobilharzia brevis

Status: 1.
Dist: PAL—Japan
Hab: Abortive schistosomule in skin
Hosts: 2. Aquatic birds
1. Snails: *Lymnaea* spp.
Trans: Cercaria actively invades skin
Zoo stat: 1
Coombs and Crompton 1991

Trichobilharzia maegraithi

Status: 1.
Dist: PAL—Thailand
Hab: Abortive schistosomule in skin
Hosts: 2. Aquatic birds
1.
Trans: Cercaria actively invades skin
Zoo stat: 1
Muller 2002

Trichobilharzia ocellata

Status: 1.
Dist: Cosmopolitan
Hab: Abortive schistosomule in skin
Hosts: 2. Birds: Anatidae
1. Snails: *Lymnaea* spp.
Trans: Cercaria actively invades skin
Zoo stat: 1
Coombs and Crompton 1991

Trichobilharzia stagnicolae

Status: 1.
Dist: NEA
Hab: Abortive schistosomule in skin
Hosts: 2. Birds: Anatidae
1. Snails: *Lymnaea* spp.
Trans: Cercaria actively invades skin
Zoo stat: 1
Coombs and Crompton 1991

ECHINOSTOMATIDAE

Acanthoparyphium kurogamo

Status: 1. Five cases detected, 1998
Dist: PAL—Korea
Hab: Adult in small intestine
Hosts: 3. Shorebirds
2.
1.
Trans:
Zoo stat: 2
Chai *et al.* 1998

Acanthoparyphium tyosenense

Status: 1. First ten cases, 2001
Dist: PAL—Korea
Hab: Adult in small intestine
Hosts: 3. Birds
2. Bivalves: *Mactra, Solen* spp.; Gastropod: *Neverita* sp.
1.
Trans: Ingestion of brackish-water molluscs
Zoo stat: 2
Chai *et al.* 2001

Artyfechinostomum malayanum

Syn: *Artyfechinostomum mehrai*
Artyfechinostomum sufrartyfex
Paryphostomum sufrartyfex
Status: 1.
Dist: OR—India, Philippines, Thailand
Hab: Adult in small intestine
Hosts: 3. Hog, rat
2. Snails: *Bullastra, Digoniostoma* spp.
1. Snail
Trans:
Zoo stat: 2
Premvati and Pande 1974; Harinasuta *et al.* 1987

Artyfechinostomum oraoni

Status: 1.
Dist: OR—India
Hab: Adult in small intestine
Hosts: 3.
2.
1.
Trans:
Zoo stat: 2
Bandyopadhyay *et al.* 1995

Cathaemasia cabrerai

Status: 1. First record 1984
Dist: OR—Philippines
Hab: Recovered from human stool
Hosts: 3. Birds?
2.
1.
Trans:
Zoo stat: 2
Coombs and Crompton 1991

Echinochasmus fujianensis

Status: 1, but up to 7.8% in Fujian, China
Dist: PAL—China
Hab:
Hosts: 3. Pig, dog, cat
2.
1. *Bellamya* sp.
Trans: Ingestion of metacercaria in fish
Zoo stat: 2
Muller 2000, 2002

Echinochasmus japonicus

Status: 1. First mentioned in 1985
Dist: PAL—China, Japan, Korea, Taiwan
Hab:
Hosts: 3. Cat, dog, rat, night heron *Nycticorax* sp.
2. Freshwater fish and amphibians
1. Snail: *Parafossarulus* sp.
Trans: Ingestion of metacercaria in fish
Zoo stat: 2
Coombs and Crompton 1991

Echinochasmus jiufoensis

Status: 1. First case, 1988
Dist: PAL—China
Hab: Adult in intestine
Hosts: 3:
2:
1:
Trans: Ingestion of metacercaria in fish
Zoo stat: 2
Coombs and Crompton 1991

Echinochasmus liliputanus

Status: 1, but up to 23% in Anhui, China
Dist: PAL—China
Hab:
Hosts: 3. Dog
2. Freshwater fish
1.
Trans: Ingestion of metacercaria in fish
Zoo stat: 2
Muller 2000, 2002

Echinochasmus perfoliatus

Status: 1. Few human cases recorded
Dist: PAL—Widespread, with human cases in China, Japan
Hab: Adult in small intestine
Hosts: 3. Cat, dog, pig etc.
2. Freshwater fish
1. Snails: *Bythinia*, *Lymnaea*, *Parafossarulus* spp.
Trans: Ingestion of metacercaria in fish
Zoo stat: 2
Beaver *et al.* 1984; Coombs and Crompton 1991

Echinoparyphium recurvatum

Status: 1. 'Several' human cases recorded
Dist: PAL—Egypt, Taiwan;
OR—Indonesia
Hab: Adult in small intestine
Hosts: 3. Domestic birds, owl *Strix* sp.
2. Snails and bivalves: *Lymnaea*, *Planorbis*, *Corbicula* spp.
1. Snails: *Lymnaea*, *Planorbis* spp.
Trans: Ingestion of metacercaria in tadpole, frog, snail or clam
Zoo stat: 2
Beaver *et al.* 1984; Harinasuta *et al.* 1987; Coombs and Crompton 1991

Echinostoma angustitestis

Status: 1. Two cases in China
Dist: PAL—China
Hab: Small intestine
Hosts: 3.
2. Freshwater fish
1.
Trans: Ingestion of metacercaria in fish
Zoo stat: 2
Muller 2000, 2002

Echinostoma cinetorchis

Status: 1. 'Occasional'
Dist: PAL—Japan, Korea, Taiwan
Hab: Adult in small intestine
Hosts: 3. Various mammals and birds
2. Snails and amphibia
1. Snail: *Segmentina* sp.
Trans: Ingestion of metacercaria in snail, tadpole or frog
Zoo stat: 2
Beaver *et al.* 1984; Harinasuta *et al.* 1987; Coombs and Crompton 1991

Echinostoma echinatum

Syn: *Echinostoma lindoense*
Status: 3. 'Man is a favoured host.' Once common in Lindu Valley, Sulawesi, but has died out, possibly due to introduction of *Tilapia* fish
Dist: OR—Indonesia, in the Lake Lindoe area of Sulawesi
Hab: Adult in small intestine
Hosts: 3.
2. Freshwater molluscs: *Corbicula*, *Viviparus* spp.
1. Snails: *Gyraulus* spp.
Trans: Ingestion of metacercaria in freshwater molluscs
Zoo stat: 2, possibly 3
Huffman and Fried 1990; Coombs and Crompton 1991; Lloyd and Soulsby 1998

Echinostoma hortense

Status: 3. Up to 22.4% in Korea, not uncommon in northwest China
Dist: PAL—China, Japan, Korea
Hab: Adult in small intestine
Hosts: 3. Rodents, dog, cat
2. Amphibia, freshwater fish: loach *Misgurnus* sp.
1. Snails: *Lymnaea* spp.
Trans: Ingestion of metacercaria in fish or frog
Zoo stat: 2
Huffman and Fried 1990; Coombs and Crompton 1991; Muller 2002

Echinostoma ilocanum

Status: 3. Rare and sporadic, but prevalence up to 44% in Philippines
Dist: PAL—China; OR—Indonesia, Malaysia, Philippines, Thailand
Hab: Adult in small intestine
Hosts: 3. Cat, dog, rat
2. Snails: *Gyraulus*, *Pila*, *Viviparus* spp.
1. Snails: *Gyraulus*, *Hippeutis* spp.
Trans: Ingestion of metacercaria in snails
Zoo stat: 2
Beaver *et al.* 1984; Harinasuta *et al.* 1987; Coombs and Crompton 1991

Echinostoma japonicum

Status: 1.
Dist: PAL—Japan, Korea
Hab:
Hosts: 3.
2.
1.
Trans:
Zoo stat: 2
Muller 2002

Echinostoma macrorchis

Status: 1. 'Less common species'
Dist: PAL—Japan, Korea, Taiwan; OR—Indonesia
Hab: Adult in small intestine
Hosts: 3. Rodents: *Rattus*; birds: Snipe *Capella* sp.
2. Snails: *Parafossarulus*, *Segmentina*, *Viviparus* spp.
1. Snails: *Parafossarulus*, *Segmentina* spp.
Trans: Ingestion of metacercaria in snails
Zoo stat: 2
Harinasuta *et al.* 1987; Coombs and Crompton 1991

Echinostoma malayanum

Status: 3. Common in northern Thailand, but rare elsewhere
Dist: OR—Thailand, India, Indonesia, Malaysia, Philippines, Singapore; PAL—China
Hab: Adult in small intestine
Hosts: 3. Pig, rat, shrew
2. Snails: *Bellamya*, *Pila* spp., tadpole, fish
1. Snails: *Lymnaea*, *Indoplanorbis* spp.
Trans: Ingestion of metacercaria in snail, tadpole or fish
Zoo stat: 2
Beaver *et al.* 1984; Harinasuta *et al.* 1987; Huffman and Fried 1990; Coombs and Crompton 1991; Maji *et al.* 1993, Muller 2000

Echinostoma melis

Syn: *Euparyphium melis*
Echinostoma jassyense
Status: 1. 'Less common species'
Dist: PAL—Romania, China, Taiwan
Hab:
Hosts: 3.
2.
1.
Trans: Ingestion of metacercaria in snails
Zoo stat: 2
Harinasuta *et al.* 1987; Muller 2002

Echinostoma revolutum

Syn: *Echinoparyphium paraulum*
Status: 3. 'Prevalence 3–6% in Taiwan'
Dist: PAL—China, Russia, Taiwan; OR—Indonesia, Thailand
Hab: Adult in small intestine
Hosts: 3. Ducks, geese, rats
2. Molluscs: *Corbicula*, *Helisoma*, *Stagnicola*, *Viviparus* spp.
1. Snails: *Helisoma*, *Stagnicola* spp.
Trans: Ingestion of metacercaria in molluscs
Zoo stat: 2
Beaver *et al.* 1984; Harinasuta *et al.* 1987; Coombs and Crompton 1991

Episthmium caninum

Status: 1. First reported 1985
Dist: OR—Thailand
Hab: Eggs recovered in stool
Hosts: 3. Adults in birds, rarely in mammals
2. Fish
1.
Trans: Presumably by ingestion of metacercaria in fish flesh
Zoo stat: 2
Harinasuta *et al.* 1987; Coombs and Crompton 1991; Muller 2000

Himasthla muehlensi

Status: 1. A single recorded infection
Dist: Either NEO or NEA
Hab: Intestine
Hosts: 3. Bird
2. Possibly clams: *Venus* sp.
1.
Trans: (Ingestion of metacercaria in clams)
Zoo stat: 2
Beaver *et al.* 1984; Harinasuta *et al.* 1987; Coombs and Crompton 1991

Hypoderaeum conoideum

Status: 3. 'Common...in north-east Thailand'; up to 55% infected in a 1965 sample
Dist: OR—Thailand, Taiwan
Hab: Adult in small intestine
Hosts: 3. Anseriform birds
2. Snails and tadpoles
1. Snails: *Lymnaea*, *Planorbis* spp.
Trans: Ingestion of metacercaria in snail
Zoo stat: 2
Beaver *et al.* 1984; Harinasuta *et al.* 1987; Coombs and Crompton 1991

Paryphostomum bangkokensis

Status: 1.
Dist: OR—Thailand
Hab: Adult in small intestine
Hosts: 3.
2.
1.
Trans:
Zoo stat: 2
Muller 2002

FASCIOLIDAE

Fasciola gigantica

Syn: *Fasciola indica*
Status: 2. High prevalences occur in certain areas
Dist: AETH; PAL; NEA; OR
Hab: Adult in liver and biliary system
Hosts: 2. Numerous ungulates
1. Snails: *Lymnaea*, *Physopsis* spp.
Trans: Ingestion of metacercaria on aquatic vegetation
Zoo stat: 2
Malek 1980b; Beaver *et al.* 1984; Coombs and Crompton 1991; Chowdhury and Tada 2001; ICZN 2001

Fasciola hepatica

Status: 3. Widespread but sporadic
Dist: Cosmopolitan
Hab: Adult in liver and biliary system
Hosts: 2. Numerous ungulates
1. Snails: *Lymnaea* spp.
Trans: Ingestion of metacercaria on aquatic vegetation
Zoo stat: 2
Beaver *et al.* 1984; Coombs and Crompton 1991

Fasciolopsis buski

Status: 4. Common within its range; estimated 10 million cases in 1980
Dist: PAL—China, Taiwan; OR—much of east Asia
Hab: Adult in small intestine
Hosts: 2. Domestic dog, pig
1. Snails: *Hippeutis*, *Segmentina*, *Gyraulus* spp.
Trans: Ingestion of metacercaria on aquatic vegetation
Zoo stat: 3
Malek 1980a; Beaver *et al.* 1984; Harinasuta *et al.* 1987; Coombs and Crompton 1991

PHILOPHTHALMIDAE

Philophthalmus lacrymosus

Status: 1. A single human case, 1941
Dist: PAL—Yugoslavia
Hab: Adult in conjunctival sac of eye
Hosts: 2. Birds: *Larus* spp. and other fish-eating birds
1.
Trans: Direct invasion of the eye, while bathing?
Zoo stat: 2
Beaver *et al.* 1984; Coombs and Crompton 1991

There is also a single record of *Philophthalmus* sp. from Sri Lanka.

PARAMPHISTOMIDAE

Fischeroederius elongatus

Status: 1.
Dist: PAL—China
Hab:
Hosts: 3. Ruminants
2.
1.
Trans:
Zoo stat: 2
Muller 2002

GASTRODISCIDAE

Gastrodiscoides hominis

Status: 3. Locally common; 41% prevalence in one survey; a few hundred thousand cases estimated in 1980
Dist: OR—Especially Assam, India, Myanmar, Malaysia, Vietnam, Thailand; PAL—Kasakhstan, 'USSR'; NEO—Guyana
Hab: Attached to caecum
Hosts: 2. Domestic pig, primates, rodents, ungulates
1. *Helicorbis* spp.
Trans: Probably metacercaria ingested with contaminated food
Zoo stat: 2, possibly 3
Malek 1980a; Beaver *et al.* 1984; Harinasuta *et al.* 1987; Coombs and Crompton 1991

Watsonius watsoni

Status: 1. Human infection is very rare
Dist: AETH—Nigeria, Namibia
Hab: Attached to mucosa of small intestine
Hosts: 2. Various primates
1.
Trans: Metacercaria probably ingested on water plants
Zoo stat: 2
Beaver *et al.* 1984; Harinasuta *et al.* 1987; Coombs and Crompton 1991

HETEROPHYIDAE

Apophallus donicus

Status: 1. Not in Beaver *et al.* (1984)
Dist: NEA—Canada; PAL—eastern Europe
Hab: Adult in small intestine
Hosts: 3. Cat, dog, rabbit, *Vulpes*, birds including *Nycticorax* sp.
2. Freshwater fish: *Richardsonius*, *Ptychocheilus, Salmo* spp.
1. Snail: *Flumenicola* sp.
Trans: Ingestion of metacercaria in fish
Zoo stat: 2
Coombs and Crompton 1991; Muller 2002

Centrocestus armatus

Status: 1. A single natural infection reported
Dist: PAL—Korea
Hab: Adult in small intestine
Hosts: 3. Piscivorous birds
2. Various freshwater fish
1. Snail: *Semisulcospira* sp.
Trans: Ingestion of metacercaria in fish
Zoo stat: 2
Chai and Lee 2002

Centrocestus cuspidatus

Syn: *Centrocestus caninus*
Status: 1.
Dist: OR—Thailand; PAL—Egypt, Taiwan
Hab: Adult in small intestine
Hosts: 3. Black Kite *Milvus migrans*
2. Fish
1. Snails: *Melania*, *Semisulcospira* spp.
Trans: Ingestion of metacercaria in fish
Zoo stat: 2
Skrjabin 1952; Harinasuta *et al.* 1987

Centrocestus formosanus

Status: 1.
Dist: PAL—Taiwan, China; OR—India, Philippines
Hab: Adult in small intestine
Hosts: 3. Fish-eating birds, cat, dog, rat
2. Fish and amphibia
1. *Semisulcospira*, *Melania* spp.
Trans: Ingestion of metacercaria in fish
Zoo stat: 2
Harinasuta *et al.* 1987; Coombs and Crompton 1991; Muller 2000

Centrocestus kurokawai

Status: 1.
Dist: PAL—Japan
Hab:
Hosts: 3.
2. Fish
1.
Trans: Ingestion of metacercaria in fish
Zoo stat: 2
Harinasuta *et al.* 1987

Centrocestus longus

Status: 1.
Dist: PAL—Taiwan
Hab:
Hosts: 3.
2.
1.
Trans: Ingestion of metacercaria in fish
Zoo stat: 2
Harinasuta *et al.* 1987

Cryptocotyle lingua

Status: 1.
Dist: NEA—Greenland
Hab: Adult in small intestine
Hosts: 3. Cat, dog, piscivorous mammals and birds
2. Marine fish: *Gobius*, *Pleuronectes, Tautogolabrus* spp.
1.
Trans: Ingestion of metacercaria in fish
Zoo stat: 2
Harinasuta *et al.* 1987; Coombs and Crompton 1991

Diorchitrema formosanum

Status: 1. 'Has been reported from man'
Dist: PAL—Taiwan
Hab:
Hosts: 3. Cat, rat
2.
1.
Trans: Ingestion of metacercaria in raw fish
Zoo stat: 2
Beaver *et al.* 1984; Harinasuta *et al.* 1987

Diorchitrema pseudocirratum

Syn: *Diorchitrema amplicaecale*
Stellantchasmus amplicaecalis
Status: 1.
Dist: OR—Philippines, Hawaii; PAL—Egypt
Hab: Cardiac lesions
Hosts: 3. Dog, cat
2. Fish
1. Snails
Trans: Ingestion of metacercaria in raw fish
Zoo stat: 2
Skrjabin 1952; Beaver *et al.* 1984; Harinasuta *et al.* 1987

Haplorchis pleurolophocerca

Status: 1.
Dist: PAL—Egypt
Hab: Adult presumably in small intestine
Hosts: 3. Cat
2. Fish: *Gambusia affinis*
1.
Trans: Presumably ingestion of metacercaria in fish
Zoo stat: 2
Harinasuta *et al.* 1987

Haplorchis pumilio

Status: 1.
Dist: PAL—Egypt, Iran, Taiwan; OR—India, Laos, Philippines
Hab: Adult presumably in small intestine
Hosts: 3. Piscivorous birds, cat, dog
2. Various fish: *Puntius* sp.
1. Snails: *Melania* spp.
Trans: Ingestion of metacercaria in fish
Zoo stat: 2
Coombs and Crompton 1991

Haplorchis taichui

Syn: *Haplorchis microrchis*
Status: 1.
Dist: PAL—Egypt, Japan, Taiwan; OR—Bangladesh, India, Laos, Philippines
Hab: Adult presumably in small intestine
Hosts: 3. Cat, dog, cattle
2. Fish: *Puntius* spp.
1. Snails: *Melania*, *Melanoides* spp.
Trans: Ingestion of metacercaria in fish
Zoo stat: 2
Skrjabin 1952; Coombs and Crompton 1991

Haplorchis vanissimus

Status: 1.
Dist: OR—Philippines
Hab: Adult in small intestine
Hosts: 3. Piscivorous birds: *Heliastur*, *Nycticorax*, *Phalacrocorax* spp.
2.
1.
Trans: Ingestion of metacercaria in fish
Zoo stat: 2
Coombs and Crompton 1991

Haplorchis yokogawai

Status: 1.
Dist: PAL—Egypt; OR—Indonesia, India, Philippines; NEA—USA: Hawaii
Hab: Adult in small intestine
Hosts: 3. Piscivorous birds, cat, dog, cattle
2. Fish: *Cirrhina*, *Ophicephalus*, *Puntius* spp.
1. Snail: *Stenomelania* sp.
Trans: Ingestion of metacercaria in fish
Zoo stat: 2
Coombs and Crompton 1991

Heterophyes dispar

Status: 1. 'Has been found in humans in Egypt'
Dist: PAL—Egypt, Korea
Hab: Adult in small intestine
Hosts: 3. Cat, dog, *Vulpes* sp., piscivorous mammals
2. Marine fish: *Barbus*, *Mugil*, *Sciaena*, *Solea* spp.
1. Snail: *Pirenella* sp.
Trans: Ingestion of metacercaria in fish
Zoo stat: 2
Coombs and Crompton 1991; Khalil 1991; ICZN 2001

Heterophyes equalis

Status: 1. 'Has been found in humans'
Dist: PAL—Egypt
Hab:
Hosts: 3.
2.
1.
Trans:
Zoo stat: 2
Khalil 1991

Heterophyes heterophyes

Status: 3. Declining prevalence in Egypt due to declining fish-eating habits following construction of Aswan High Dam.
Dist: PAL—Egypt, France, Japan, Korea (imported)
Hab: Adult in small intestine
Hosts: 3. Cat, dog, *Vulpes*, piscivorous mammals
2. Brackish-water fish: *Aphanius*, *Mugil*, *Tilapia* spp.
1. Snails: *Pirenella*, *Cerithidea* spp.
Trans: Ingestion of metacercaria in fish
Zoo stat: 2/3
Beaver *et al.* 1984; Coombs and Crompton 1991; Khalil 1991

Heterophyes katsuradai

Status: 1.
Dist: PAL—Japan
Hab:
Hosts: 3. Cat, dog, rat
2. Fish
1. Snails
Trans:
Zoo stat: 2
Harinasuta *et al.* 1987

Heterophyes nocens

Status: 2. Common locally in Korea
Dist: PAL—Japan, Korea
Hab: Adult in small intestine
Hosts: 3. Recorded from man, dog, cat
2. Fish: *Acanthogobius*, *Lateolabrax*, *Mugil* spp.
1. Brackish-water snail: *Cerithidea* sp.
Trans: Ingestion of metacercaria in fish
Zoo stat: 2
Harinasuta *et al.* 1987; Coombs and Crompton 1991; Chai and Lee 2002

Heterophyopsis continua

Status: 1. Three human cases to 2002
Dist: PAL—Japan, Korea
Hab: Adult in small intestine
Hosts: 3. Piscivorous birds, cat, dog
2. Fish: *Acanthogobius*, *Lateolabrax*, *Mugil* spp.
1.
Trans: Ingestion of metacercaria in fish
Zoo stat: 2
Coombs and Crompton 1991; Chai and Lee (in press)

Metagonimus minutus

Status: 1.
Dist: PAL—China, Taiwan
Hab: Adult in small intestine
Hosts: 3. Cat
2. Fish
1.
Trans: Ingestion of metacercaria in fish
Zoo stat: 2
Coombs and Crompton 1991

Metagonimus miyatai

Status: 1., but high prevalence locally
Dist: PAL—Japan, Korea
Hab: Adult in small intestine
Hosts: 3. Kite, dog, fox
2. Sweetfish *Plecoglossus altivelis,* and other freshwater fish
1. Snail: *Semisulcospira* sp.
Trans: (Ingestion of metacercaria in fish)
Zoo stat: 2
Chai and Lee 2002

Metagonimus takahashii

Status: 1. Seven cases to 1994
Dist: PAL—Japan, Korea
Hab: Adult in small intestine
Hosts: 3.
2. Freshwater fish: *Carassius carassius* etc.
1.
Trans: Ingestion of metacercaria in fish
Zoo stat: 2
Muller 2002; Chai and Lee 2002

Metagonimus yokogawai

Syn: *Heterophyes yokogawai*
Status: 3. 'Considerable percentage infected in highly endemic areas'; estimated hundreds of thousands of cases annually in 1980
Dist: PAL—China, Israel, Japan, Korea, Russian Federation: Siberia, Spain, Taiwan; OR—Indonesia, Philippines
Hab: Adult in small intestine
Hosts: 3. Cat, dog, rat
2. Cyprinid fish: *Plecoglossus* sp.
1. Snails: *Semisulcospira* spp.
Trans: Ingestion of metacercaria in fish
Zoo stat: 2
Malek 1980a; Beaver *et al.* 1984; Harinasuta *et al.* 1987; Coombs and Crompton 1991

Phagicola longa

Syn: *Ascocotyle longa*
Status: 1.
Dist: NEA—Brazil, Mexico
Hab: Adult in small intestine
Hosts: 3. Piscivorous birds, dog
2. Mullet *Mugil* sp.
1.
Trans: Ingestion of metacercaria in fish
Zoo stat: 2
Muller 2000, 2002

Procerovum calderoni

Status: 1.
Dist: OR—Philippines; PAL—China, Egypt
Hab: Adult in small intestine
Hosts: 3. Cat, dog
2. Fish: *Butis, Chanos, Mugil, Ophicephalus, Platycephalus* spp.
1. Snails: *Melania, Thiara* spp.
Trans: Ingestion of metacercaria in fish
Zoo stat: 2
Harinasuta *et al.* 1987; Coombs and Crompton 1991

Pygidiopsis summa

Status: 1. Sporadic cases in coastal areas of Korea
Dist: PAL—Japan, Korea
Hab: Adult in small intestine; up to 4000 worms recovered
Hosts: 3. Piscivorous birds, cat
2. Mullet, goby
1. Brackish-water snail, *Cerithidea, Tympanotus* spp.
Trans: Ingestion of metacercaria in fish
Zoo stat: 2
Beaver *et al.* 1984; Coombs and Crompton 1991; Chai and Lee 2002

Stellantchasmus falcatus

Status: 1. 'Established in Hawaii'
Dist: NEA—USA: Hawaii; PAL—Japan, Korea; OR—Laos, Philippines, Thailand
Hab: Intestine
Hosts: 3. Cat, dog, rat, piscivorous birds: *Colymbus* sp.
2. Fish: *Acanthogobius*, *Liza*, *Mugil*
1. Snails: *Stenomelania*, *Thiara* spp.
Trans: Ingestion of metacercaria on fish fins
Zoo stat: 2
Beaver *et al.* 1984; Coombs and Crompton 1991

Stictodora fuscata

Status: 1. Fourteen cases to 2002 in Korea
Dist: PAL—Korea
Hab: Alimentary tract
Hosts: 3. Piscivorous birds
2. Mullet, goby
1.
Trans: Ingestion of metacercaria in fish
Zoo stat: 2
Chai and Lee 2002

Stictodora lari

Status: 1. First case in 1998
Dist: PAL— Korea
Hab: Small intestine
Hosts: 3. Piscivorous birds
2. Brackish-water fish
1. Brackish-water snail: *Velacuminatus* sp. (in Australia)
Trans: (Ingestion of metacercaria in fish)
Zoo stat: 2
Chai and Lee 2002

Stictodora manilensis

Status: 1.
Dist: PAL—Korea
Hab: Small intestine
Hosts: 3. Piscivorous birds, dog
2. Mullet, goby
1.
Trans: Ingestion of metacercaria in fish
Zoo stat: 2
Muller 2000

OPISTHORCHIIDAE

Clonorchis sinensis

Syn: *Opisthorchis sinensis*
Status: 4. Five hundred autopsy cases mentioned by Beaver *et al.* (1984); up to 28% prevalence in Vietnam; 19 million cases estimated, in 1980
Dist: PAL—China, Japan, Korea, Taiwan; OR—Vietnam
Hab: Adult in liver, biliary system, pancreatic duct
Hosts: 3. Cat, dog, pig, piscivorous carnivores: *Martes*, *Meles*, *Mustela* spp.
2. Numerous cyprinid fish
1. Snails: *Bulimus*, *Parafossarulus* spp.
Trans: Ingestion of metacercaria in fish
Zoo stat: 3, but 'can be regarded as primarily a human parasite'
Malek 1980a; Beaver *et al.* 1984; Coombs and Crompton 1991; Lloyd and Soulsby 1998; Muller 2000

Metorchis albidus

Status: 1.
Dist: NEA—USA: Alaska; PAL—Iran, France, Turkey, 'USSR'
Hab: Adult presumably in biliary system
Hosts: 3. Normally in fish-eating mammals
2. Various freshwater fish
1. *Bithynia* sp.
Trans: Ingestion of metacercaria in fish
Zoo stat: 2
Coombs and Crompton 1991; Muller 2000

Metorchis conjunctus

Status: 1.
Dist: NEA—Canada, Greenland
Hab: Presumably bile ducts
Hosts: 3. Normally in fish-eating mammals
2. Fish: *Catostomus* sp.
1. Snail: *Amnicola* sp.
Trans: Ingestion of metacercaria in fish
Zoo stat: 2
Coombs and Crompton 1991; Muller 2000

Opisthorchis felineus

Status: 3. Endemic foci in deltas and reservoirs; reached 95% prevalence in Ob'-Irtysh basin, Siberia; estimated 1.1 million cases in 1980
Dist: PAL—Central, eastern and southern Europe, Siberia
Hab: Adult in liver, biliary system and lung
Hosts: 3. Cat, dog, pig
2. Cyprinid fish: *Abramis*, *Barbus*, *Tinca*
1.
Trans: Ingestion of metacercaria in fish
Zoo stat: 3?
Malek 1980a; Beaver *et al.* 1984; Coombs and Crompton 1991; Lloyd and Soulsby 1998

Opisthorchis guayaquilensis

Syn: *Amphimerus pseudofelineus*
Status: 2. Common in one isolated focus
Dist: NEO—Brazil, Ecuador, Panama
Hab: Bile duct, pancreatic duct
Hosts: 3. Cat, coyote, dog
2.
1.
Trans: Ingestion of metacercaria in fish
Zoo stat: 2
Beaver *et al.* 1984; Coombs and Crompton 1991

Opisthorchis viverrini

Status: 3. Up to 46% prevalence in north-eastern Thailand; estimated 3.5 million cases in 1980
Dist: OR—Laos, Thailand
Hab: Adult in biliary system
Hosts: 3. Cat, dog, Fishing Cat *Felis viverrinus*, various other piscivorous mammals
2. Freshwater fish: *Cyclocheilichthys*, *Hampala*, *Puntius* spp.
1. Snails: *Bithynia* spp.
Trans: Ingestion of metacercaria in fish
Zoo stat: 2
Malek 1980a; Beaver *et al.* 1984; Coombs and Crompton 1991

Pseudamphistomum aethiopicum

Status: 1. A single record, from 1941
Dist: AETH—Begemdir, Ethiopia
Hab: In cyst-like nodules on internal wall of small intestine
Hosts: 3. Only known from man
2.
1.
Trans:
Zoo stat: 1
Cacciapuoti 1947; Coombs and Crompton 1991

Pseudamphistomum truncatum

Status: 1. First human case recorded 1982
Dist: PAL—Germany, Italy, Portugal, Russia
Hab: Adult in bile ducts
Hosts: 3. Cat, dog, *Vulpes*, *Mustela*, *Phoca* spp.
2. Freshwater fish: *Abramis*, *Carassius* spp.
1.
Trans: Ingestion of metacercaria in fish
Zoo stat: 2
Coombs and Crompton 1991

MICROPHALLIDAE

Carneophallus brevicaeca

Syn: *Spelotrema brevicaeca*
Status: 1. 'Several records'
Dist: OR—Philippines
Hab: Alimentary tract; eggs—from ectopic adults—in various tissues
Hosts: 3. Primates: *Macaca* sp.; piscivorous birds: *Sterna* sp.
2. Shrimp: *Macrobrachium* sp.; crab: *Cararius* sp.
1.
Trans: Ingestion of metacercaria in shrimp
Zoo stat: 2
Coombs and Crompton 1991

DICROCOELIIDAE

Dicrocoelium dendriticum

Status: 1. 'Numerous records; most but not all are of pseudoparasitism'
Dist: Cosmopolitan
Hab: Adult in biliary system
Hosts: 3. Ungulate mammals: sheep etc.
2. Various ants
1. Many terrestrial snail species
Trans: Ingestion of metacercaria in ant or on contaminated vegetation
Zoo stat: 2
Beaver *et al.* 1984; Coombs and Crompton 1991; ICZN 2001

Dicrocoelium hospes

Status: 1. Two valid records
Dist: AETH—Ethiopia, Ghana, Kenya, Sierra Leone
Hab: Adult in biliary system
Hosts: 3. Ungulate mammals
2. Ants: *Crematogaster*, *Dorylus* spp.
1. Snails: *Achatina*, *Limicolaria* spp.
Trans: Ingestion of metacercaria in ant
Zoo stat: 2
Beaver *et al.* 1984; Coombs and Crompton 1991

Eurytrema pancreaticum

Status: 1. At least eight records
Dist: PAL—China, Japan; OR; NEO
Hab: Adult in pancreatic duct
Hosts: 3. Mainly sheep, also various ungulates and primates: *Macaca* sp.
2. Grasshopper: *Conocephalus* sp.
1. Terrestrial snail: *Bradybaena* sp.
Trans: Ingestion of metacercaria in grasshopper
Zoo stat: 2
Beaver *et al.* 1984; Coombs and Crompton 1991; Smyth 1995

LECITHODENDRIIDAE

Phaneropsolus bonnei

Status: 3. 'Found in 15 of 24 autopsies in northern Thailand'; sometimes reaches 40%
Dist: OR—Thailand, Laos, Indonesia
Hab: Adult in small intestine
Hosts: 3. Primates: *Macaca*, *Nycticebus* spp.; bat
2. Dragonfly naiad
1.
Trans: Consumption of small fish which have eaten dragonfly naiads
Zoo stat: 2
Harinasuta *et al.* 1987; Coombs and Crompton 1991

Prosthodendrium molenkampi

Status: 3. 'Found in 14 of 24 autopsies in northern Thailand'
Dist: OR—Thailand, Indonesia
Hab: Adult in small intestine
Hosts: 3. Rodents, bats, primates
2. Dragonfly naiad
1.
Trans: Consumption of small fish which have eaten dragonfly naiads
Zoo stat: 2
Harinasuta *et al.* 1987; Coombs and Crompton 1991

PLAGIORCHIIDAE

Plagiorchis harinasutai

Status: 1. First described in 1989
Dist: OR—Thailand
Hab: Adult in small intestine
Hosts: 3. Found in rats
2.
1.
Trans:
Zoo stat: 2
Coombs and Crompton 1991

Plagiorchis javensis

Status: 1. A single record
Dist: OR—Indonesia
Hab: Adult in small intestine
Hosts: 3. Bird, bat
2. Larval insect
1.
Trans:
Zoo stat: 2
Beaver *et al.* 1984; Harinasuta *et al.* 1987; Coombs and Crompton 1991

Plagiorchis muris

Status: 1. Very few records
Dist: PAL—Japan, Korea
Hab: Adult in small intestine
Hosts: 3. Rodents, dog, sheep, birds
2. Snail: *Lymnaea* sp. and midge: *Chironomus* sp.
1. Snails: *Lymnaea* spp.
Trans:
Zoo stat: 2
Beaver *et al.* 1984; Coombs and Crompton 1991; Muller 2000

Plagiorchis philippinensis

Status: 1. One record apparently
Dist: OR—Philippines: Ilocano
Hab: Adult in small intestine
Hosts: 3. Bird, rat
2. Insects?
1.
Trans:
Zoo stat: 2
Beaver *et al.* 1984; Harinasuta *et al.* 1987; Coombs and Crompton 1991

ACHILLURBAINIIDAE

Achillurbainia congolensis

Syn: *Poikilorchis congolensis*
Status: 1. Eight human cases
Dist: AETH—Central Africa, Nigeria; OR—Sarawak
Hab: Adult (usually only eggs recovered) in retro-auricular nodules
Hosts: 3. Unknown, likely to be a rodent
2.
1.
Trans:
Zoo stat: 1
Beaver *et al.* 1984; Coombs and Crompton 1991; Khalil 1991; Blair *et al.* 1999

Achillurbainia nouveli

Status: 1. One human record
Dist: PAL—China
Hab: Abortive adult in retro-auricular nodule
Hosts: 3. Leopard
2. Crab: *Paratelphusa* sp.
1.
Trans: Ingestion of metacercaria in crab
Zoo stat: 1
Beaver *et al.* 1984; Coombs and Crompton 1991

Achillurbainia recondita

Status: 1. Very few human cases
Dist: NEA—USA; NEO—Brazil, Honduras
Hab: Abortive adult in omentum and other peritoneal surfaces
Hosts: 3. *Didelphis* sp.
2.
1.
Trans:
Zoo stat: 1
Beaver *et al.* 1984; Coombs and Crompton 1991

PARAGONIMIDAE

Paragonimus africanus

Status: 3. Thirty of 265 sputums positive in Cameroon; locally common in Nigeria
Dist: AETH—Cameroon, Equatorial Guinea, Nigeria
Hab: Adult in lungs
Hosts: 3. Carnivora: dog, mongoose *Crossarchus*, civet *Viverra*; *Malacomys* and *Crocidura*
2. Freshwater crabs: *Sudanonautes* spp.
1.
Trans: Ingestion of metacercaria in crab
Zoo stat: 2
Blair *et al.* 1999

Paragonimus heterotremus

Status: 3. May be important locally
Dist: OR—China, Laos, Thailand, Vietnam
Hab: Adult in lungs
Hosts: 3. Numerous primates, carnivores and rodents
2. Freshwater crabs: Parathelphusidae, Potamidae
1. Snails: *Assiminea, Neotricula, Oncomelania* spp.
Trans: Ingestion of metacercaria in crab
Zoo stat: 2
Blair *et al.* 1999

Paragonimus kellicotti

Status: 1. One human record
Dist: NEA—USA, Canada
Hab: Adult in lungs
Hosts: 3. Cat, dog, pig, opossum, mink, raccoon
2. Crayfish: Astacidae
1. Snail: *Pomatiopsis* sp.
Trans: Ingestion of metacercaria in crayfish
Zoo stat: 2
Blair *et al.* 1999

Paragonimus mexicanus

Status: 1. Four cases cited
Dist: NEO—Mexico, Peru, Ecuador, Costa Rica, Panama, Guatemala
Hab: Adult in lungs
Hosts: 3. Didelphidae, Canidae, Felidae, Mustelidae, Procyonidae, Suidae
2. Crabs: Pseudothelphusidae, Trichodactylidae
1. Snail: *Aroapyrgus* sp.
Trans: Ingestion of metacercaria in crab
Zoo stat: 2
Blair *et al.* 1999

Paragonimus miyazakii

Status: 2. 'Many recent cases'
Dist: PAL—Japan
Hab: Does not mature in man
Hosts: 3. Carnivores: dog, Mustelidae
2. Crab: *Geothelphusa* sp.
1. Snails: *Bythinella*, *Oncomelania* spp.
Trans: Ingestion of metacercaria in crab
Zoo stat: 1
Blair *et al.* 1999; Muller 2002

Paragonimus ohirai

Syn: *Paragonimus iloktsuenensis*
Paragonimus sadoensis
Status: 1. First case in 1988
Dist: PAL—China, Japan, Korea, Taiwan
Hab: Adult in lungs
Hosts: 3. *Nyctereutes proyonoides,* cat, Mustelidae, *Rattus* spp.
2. Crabs: Grapsidae, Potamidae
1. Snail: *Angustassiminea*, *Oncomelania* spp.
Trans:
Zoo stat: 2
Blair *et al.* 1999

Paragonimus skrjabini

Syn: *Paragonimus hueitungensis*
Status: 2. One study involved 119 cases
Dist: PAL—China; OR—Thailand
Hab: Immature in lungs, subcutaneous and cerebral tissue
Hosts: 3. Canidae, Felidae, Mustelidae, Viverridae; *Rattus, Hystrix* spp.
2. Crabs: Potamidae; amphibia
1. Snails: Assimineidae, Pomatiopside
Trans: Ingestion of metacercaria in crab
Zoo stat: 1; 'cannot mature in humans'
Blair *et al.* 1999

Paragonimus uterobilateralis

Status: 3. An important infection in Nigeria
Dist: AETH—Cameroon, Gabon, Liberia, Nigeria,
Hab: Adult in lungs
Hosts: 3. Carnivores: dog, *Atilax*, *Lutra, Civettictis* spp.
2. Crabs: *Liberonautes*, *Sudanonautes* spp.
1. Snails: probably *Afropomus*, *Potadoma* spp.
Trans: Ingestion of metacercaria in crab
Zoo stat: 2
Blair *et al.* 1999

Paragonimus westermanii

Syn: *Paragonimus philippinensis*
Paragonimus pulmonalis
Paragonimus westermani
Status: 4. Very common in many parts of south-east Asia; estimated three million cases in 1980
Dist: PAL—China, Japan, Korea, Taiwan, 'USSR'; OR—Cambodia, India, Indonesia, Laos, Malaysia, Myanmar, Nepal, Pakistan, Philippines, Sri Lanka, Thailand, Vietnam; AUS—New Guinea
Hab: Adult in lungs and various ectopic sites
Hosts: 3. Cat, dog, pig, wolf *Canis lupus*, *Herpestes*, *Viverra* spp.
2P. Wild boar
2. Numerous freshwater crabs and crayfish
1. Snails: e.g. *Semisulcospira*, *Brotia*, *Thiara* spp.
Trans: Ingestion of metacercaria in crabs and crayfish
Zoo stat: 3. 'Human pollution is less important than ... reservoir hosts'
Blair *et al.* 1999

TROGLOTREMATIDAE

Nanophyetus salmincola salmincola

Syn: *Troglotrema salmincola*
Status: 1. Reported from man occasionally in north-western USA
Dist: NEA—USA
Hab: Adult in small intestine
Hosts: 3. Carnivora: *Procyon*, spotted skunk; piscivorous birds
2. Salmonid fish: *Oncorhynchus* spp.
1. Snails: *Goniobasis*, *Oxytrema* spp.
Trans: Ingestion of metacercaria in fish
Zoo stat: 2
Beaver *et al.* 1984; Coombs and Crompton 1991; Lloyd and Soulsby 1998

Nanophyetus salmincola schikhobalowi

Syn: *Troglotrema salmincola*
Status: 2.'Commonly reported from man in eastern Russia'
Dist: PAL—Russian Federation: Siberia
Hab: Adult in small intestine
Hosts: 3. Various Carnivora
2. Various fish
1. Snails: *Semisulcospira* spp.
Trans: Ingestion of metacercaria in fish
Zoo stat: 2
Beaver *et al.* 1984; Coombs and Crompton 1991; Lloyd and Soulsby 1998

Part 3: Cestoda

DIPHYLLOBOTHRIIDAE

Diphyllobothrium cameroni

'*cordatum*' group

Status: 1. First human record in 1981
Dist: PAL—Japan
Hab: Adult in small intestine
Hosts: 3. Seals
2. Marine fish
1. (Copepod)
Trans: (Ingestion of plerocercoid in fish flesh)
Zoo stat: 2

Andersen *et al.* 1987; Coombs and Crompton 1991

Diphyllobothrium cordatum

Status: 1.
Dist: NEA—Alaska
Hab: Adult in small intestine
Hosts: 3. Seals and toothed whales, mainly the former
2. Marine fish
1.
Trans: (Ingestion of plerocercoid in fish flesh)
Zoo stat: 2

Andersen *et al.* 1987; Coombs and Crompton 1991

Diphyllobothrium dalliae

Status: 2. A common parasite of man in western Alaska
Dist: NEA
Hab: Adult in small intestine
Hosts: 3. Dog, Arctic fox, birds: Laridae
2. Freshwater fish
1.
Trans: (Ingestion of plerocercoid in fish flesh)
Zoo stat: 2

Rausch and Hilliard 1970; Coombs and Crompton 1991

Diphyllobothrium dendriticum

Syn: *Diphyllobothrium giljacicum*?
Diphyllobothrium minus
Diphyllobothrium nenzi
Diphyllobothrium skrjabini?
Diphyllobothrium ursi pro parte
Diphyllobothrium tungussicum
Status: 1.
Dist: PAL; NEA
Hab: Adult in small intestine
Hosts: 3. Main hosts are birds: Laridae; also piscivorous mammals
2. Freshwater fish
1. Copepods
Trans: (Ingestion of plerocercoid in fish flesh)
Zoo stat: 2

Andersen *et al.* 1987; Coombs and Crompton 1991; Lloyd 1998a

Diphyllobothrium elegans

Status: 1.
Dist: PAL—Japan
Hab: Adult in small intestine
Hosts: 3. Seals
2. Marine fish
1.
Trans: (Ingestion of plerocercoid in fish flesh)
Zoo stat: 2

Andersen *et al.* 1987; Coombs and Crompton 1991

Diphyllobothrium hians

'*cordatum*' group

Status: 1. First human record in 1988
Dist: PAL
Hab: Adult in small intestine
Hosts: 3. Seals
2. Marine fish
1.
Trans:
Zoo stat: 2

Andersen *et al.* 1987; Coombs and Crompton 1991

Diphyllobothrium klebanovskii

Status: 2. Prevalence up to 4%
Dist: PAL
Hab: Adult in small intestine
Hosts: 3. Main host is brown bear
2. Anadromous fish
1.
Trans:
Zoo stat: 2
Andersen *et al.* 1987; Coombs and Crompton 1991; Lloyd 1998a

Diphyllobothrium lanceolatum

'*cordatum*' group
Status: 1.
Dist: NEA
Hab:
Hosts: 3. Seals, dog
2. Marine fish
1.
Trans:
Zoo stat: 2
Andersen *et al.* 1987; Coombs and Crompton 1991

Diphyllobothrium latum

Status: 3. Was once abundant in much of Europe, now much less so
Dist: PAL; NEA
Hab: Adult in small intestine
Hosts: 3. Seals, dogs, bears
2. Freshwater fish
1. Copepods
Trans: Ingestion of plerocercoid in fish flesh
Zoo stat: 3
Andersen *et al.* 1987; Coombs and Crompton 1991

Diphyllobothrium nihonkaiense

Status: 1.
Dist: PAL—Japan
Hab: Adult in small intestine
Hosts: 3. Only known from man
2. Anadromous fish
1.
Trans:
Zoo stat: 2
Andersen *et al.* 1987; Coombs and Crompton 1991

Diphyllobothrium orcini

Status: 1.
Dist:
Hab:
Hosts:
Trans:
Zoo stat: 2
Andreassen 1998; Lloyd 1998a

Diphyllobothrium pacificum

'*elegans*' group
Status: 2. The 335 cases reported in Peru between 1966 and 1986 were almost certainly infected with this species
Dist: PAL; NEO—Pacific rim
Hab: Adult in small intestine
Hosts: 3. Eared seals
2. Marine fish
1.
Trans:
Zoo stat: 2
Lumbreras-Cruz *et al.* 1986; Andersen *et al.* 1987; Coombs and Crompton 1991

Diphyllobothrium scoticum

Status: 1. First human record in 1988
Dist: PAL—Japan
Hab: Adult in small intestine
Hosts: 3. Leopard seal, sea lion (but these are not PAL!)
2.
1.
Trans:
Zoo stat: 2
Coombs and Crompton 1991

Diphyllobothrium stemmacephalum

Syn: *Diphyllobothrium yonagoensis*
Status: 1.
Dist:
Hab: Adult in small intestine
Hosts: 3. Cetaceans
2. Marine fish
1.
Trans:
Zoo stat: 2
Andersen *et al.* 1987; Coombs and Crompton 1991

Diplogonoporus brauni

Status: 1. Two human cases
Dist: PAL—Romania
Hab: Adult in small intestine
Hosts: 3. Probably piscivorous bird
2.
1.
Trans:
Zoo stat: 2
Beaver *et al.* 1984; Coombs and Crompton 1991

Diplogonoporus fukuokaensis

Status: 1. First human case in 1970
Dist: PAL—Japan
Hab: Adult in small intestine
Hosts: 3.
2.
1.
Trans:
Zoo stat: 2
Coombs and Crompton 1991

Diplogonoporus grandis

Syn: *Diplogonoporus balaenopterae*
Status: 2. Well known as a human parasite in Japan: >200 cases recorded
Dist: PAL—Japan, Korea
Hab: Adult in small intestine
Hosts: 3. Dog, whales: *Balaenoptera* spp.
2. Marine fish
1. Copepods
Trans: (Ingestion of plerocercoid in fish)
Zoo stat: 2
Beaver *et al.* 1984; Coombs and Crompton 1991; Kino *et al.* 2002

Ligula intestinalis

Status: 1.
Dist: PAL—Poland, Romania
Hab: Adult in small intestine
Hosts: 3. Piscivorous birds
2. Freshwater fish: Cyprinidae
1. Copepods: *Cyclops*, *Diaptomus*, *Mesocyclops* spp.
Trans: (Ingestion of plerocercoid in fish)
Zoo stat: 2
Coombs and Crompton 1991; ICZN 2002

Pyramicocephalus anthocephalus

Status: 1.
Dist: NEA—Greenland, USA: Alaska
Hab: Presumably adult in small intestine
Hosts: 3. Dog, piscivorous mammals: *Phoca barbata*
2. Marine fish: *Eliginus*, *Megalocottus* spp.
1.
Trans: (Ingestion of plerocercoid in fish)
Zoo stat: 2
Coombs and Crompton 1991

Schistocephalus solidus

Status: 1.
Dist: NEA—USA: Alaska
Hab: Adult in small intestine
Hosts: 3. Dog, piscivorous birds
2. Freshwater fish: *Gasterosteus* sp.
1. Copepod: *Cyclops* sp.
Trans: (Ingestion of plerocercoid in fish)
Zoo stat: 2
Coombs and Crompton 1991

Spirometra erinaceieuropaei

Syn: *Spirometra erinacei*
Spirometra mansoni
Status: 1.
Dist: PAL—Europe, East Asia; OR—South-east Asia
Hab: Adult in small intestine
Hosts: 3. Carnivores, including dogs and cats
2. Freshwater fish, amphibia, snakes
1. Copepods
Trans: Ingestion of procercoid in copepod, of plerocercoid in amphibia or snake, or migration of plerocercoid from frog flesh
Zoo stat: 2
Beaver *et al.* 1984; Coombs and Crompton 1991

Spirometra houghtoni

Status: 1. Two human cases
Dist: PAL—China
Hab: Adult in intestine; sparganum in tissues
Hosts: 3. Cat, dog
2.
1.
Trans:
Zoo stat: 2
Beaver *et al.* 1984

Spirometra mansonoides

Status: 1.
Dist: NEA
Hab: Sparganum larva in subcutaneous tissue etc.
Hosts: 3. Cat, dog, racoon etc.
2. Frogs, snakes
1. Copepods
Trans:
Zoo stat: 1
Coombs and Crompton 1991

Spirometra theileri

Status: 2.
Dist: AETH
Hab: Sparganum larva in subcutaneous tissue
Hosts: 3. Carnivora including *Crocuta crocuta*
2. Herbivorous mammals
1. Copepods
Trans:
Zoo stat: 1
Coombs and Crompton 1991

ANOPLOCEPHALIDAE

The identification of the Anoplocephalan and Davaineid cestodes is extremely difficult. Their taxonomy and nomenclature are both rather confused. Human cases are rare and sporadic, and have probably been misidentified and misnamed. We have tried to rationalise the diverse reports.

Bertiella mucronata

Status: 1. Seven human cases on record
Dist: NEO—Argentina, Brazil, Cuba, Paraguay
Hab: Adult in small intestine
Hosts: 2. Primates: *Alouatta* spp.
1. Oribatid mites
Trans: Ingestion of cysticercoid in mite
Zoo stat: 2
Beaver *et al.* 1984; Coombs and Crompton 1991; Denegri and Perez-Serrano 1997

Bertiella studeri

Syn: *Bertiella satyri*
Status: 1. At least 44 cases of human infection, mostly from villages where monkeys are kept in captivity
Dist: AETH; OR—Widespread but sporadic
Hab: Adult in small intestine
Hosts: 2. Primates: *Pongo*, *Macaca* spp.
1. Oribatid mites: *Scheloribates* sp.
Trans: Ingestion of cysticercoid in mite
Zoo stat: 2
Beaver *et al.* 1984; Coombs and Crompton 1991; Denegri and Perez-Serrano 1997; Lloyd 1998a

Inermicapsifer cubensis

Syn: Possibly synonymous with *Inermicapsifer madagascariensis*
Status: 2. 'Often found in man in Cuba'
Dist: NEO—Cuba, Venezuela
Hab: Adult in small intestine
Hosts: 2. Rodents
1.
Trans:
Zoo stat: 3
Beaver *et al.* 1984; Lloyd 1998a

Inermicapsifer madagascariensis

Syn: *Raillietina madagascariensis*
Possibly synonymous with *Inermicapsifer cubensis*
Status: 2. Sporadic cases in Africa
Dist: AETH—Comoros, Congo, Kenya, Madagascar, South Africa, Zambia, Zimbabwe; OR—Indonesia, Thailand
Hab: Adult in small intestine
Hosts: 2. Rodents: *Praomys* spp.
1.
Trans:
Zoo stat: 2
Beaver *et al.* 1984; Coombs and Crompton 1991; Lloyd 1998a

Killigrewia sp.

Status: 1.
Dist: OR—India
Hab: Adult in small intestine
Hosts: 2. Birds: Columbiformes
1.
Trans:
Zoo stat: 2
Chowdhury and Tada 2001

Mathevotaenia symmetrica

Status: 1. First human record in 1986
Dist: OR—Thailand
Hab: Adult in small intestine
Hosts: 2. Rodents, including *Mus* sp.
1. Insects: *Tribolium*, *Plodia* spp.
Trans: Ingestion of cysticercoid in insect
Zoo stat: 2
Coombs and Crompton 1991

Moniezia expansa

Status: 1. A single human case on record
Dist: PAL—'USSR'
Hab: Adult in small intestine
Hosts: 2. Ungulates: cattle, sheep, goats etc.
1. Oribatid mites: *Galumna*, *Scheloribates* spp.
Trans: Ingestion of cysticercoid in mite
Zoo stat: 2
Coombs and Crompton 1991; ICZN 2001

DAVAINEIDAE

Buginetta alouattae

Status: 1.
Dist: NEO
Hab: Adult in small intestine
Hosts: 2. Monkeys
1.
Trans:
Zoo stat: 2
Muller 2002

Raillietina celebensis

Status: 1.
Dist: PAL—China, Japan, Taiwan; OR—Philippines; AUS—Australia, Tahiti
Hab: Adult in small intestine
Hosts: 2. Rodents: *Mus*, *Rattus* spp.
1. Ant: *Cardiocondyle* sp.
Trans: Ingestion of cysticercoid in ant
Zoo stat: 2
Coombs and Crompton 1991

Raillietina demerariensis

Status: 2. Prevalence reaches 5% in rural areas near Quito, Ecuador
Dist: NEO—Guyana, Ecuador, Cuba
Hab: Adult in small intestine
Hosts: 2. Rodents
1. Cockroach?
Trans: Ingestion of cysticercoid in insect
Zoo stat: 2
Beaver *et al.* 1984

DILEPIDIDAE

Dipylidium caninum

Status: 1. About 200 case reports from around the world
Dist: Cosmopolitan except AUS
Hab: Adult in small intestine
Hosts: 2. Cat, dog, other Carnivora
1. Fleas: *Ctenocephalides*, *Pulex* spp.
Trans: Ingestion of cysticercoid in flea
Zoo stat: 2
Coombs and Crompton 1991; Lloyd 1998a; ICZN 2001

HYMENOLEPIDIDAE

Drepanidotaenia lanceolata

Status: 1.
Dist: NEA—USA
Hab: Adult in small intestine?
Hosts: 2. Anseriform birds
1. Copepods: *Cyclops*, *Diaptomus* spp.
Trans: Ingestion of cysticercoid in copepod
Zoo stat: 2
Coombs and Crompton 1991

Hymenolepis diminuta

Status: 2. More than 200 human cases
Dist: Cosmopolitan
Hab: Adult in small intestine
Hosts: 2. Rodents: rat, mouse, *Praomys* sp.
1. A wide variety of insects and other arthropods
Trans: Ingestion of cysticercoid in intermediate host
Zoo stat: 2
Beaver *et al.* 1984; Coombs and Crompton 1991; ICZN 2001

Rodentolepis nana

Syn: *Hymenolepis fraterna*
Hymenolepis nana
Vampirolepis nana
Status: 4. Reaches 9% prevalence locally in Argentina; highest prevalence in areas of high temperature and low rainfall
Dist: Cosmopolitan
Hab: Cysticercoid and adult worm in small intestine
Hosts: 2. Rodents?
1. Various insects (optional)
Trans: Ingestion of egg by faeco–oral contamination (direct); ingestion of cysticercoid in insect
Zoo stat: 4/3 depending on taxonomy
Beaver *et al.* 1984; Coombs and Crompton 1991

It is unclear whether the human parasites are identical with those in rodents.

MESOCESTOIDAE

Mesocestoides lineatus

Status: 1. At least ten cases recorded in Japan and two in Korea
Dist: PAL—China, Japan, Korea
Hab: Adult in small intestine
Hosts: 3. Carnivora, including foxes
2. Amphibia, reptiles, small mammals
1.
Trans: Ingestion of tetrathyridium in intermediate host 2
Zoo stat: 2
Beaver *et al.* 1984; Coombs and Crompton 1991

Mesocestoides variabilis

Status: 1. Seven human cases on record
Dist: NEA—Greenland, USA
Hab: Adult in small intestine
Hosts: 3. Carnivora
2. Birds, snakes, frogs, rodents
1.
Trans: Ingestion of tetrathyridium in intermediate host 2.
Zoo stat: 2
Beaver *et al.* 1984; Coombs and Crompton 1991

TAENIIDAE

Echinococcus granulosus

Status: 3. Human infection is widespread but common only in exceptional circumstances
Dist: Cosmopolitan
Hab: Hydatid cyst in liver, lungs etc.
Hosts: 2. Dog and other Carnivora
1. Various, mainly ungulates
Trans: Ingestion of egg by contamination
Zoo stat: 1
Beaver *et al.* 1984; Coombs and Crompton 1991; ICZN 2001

Echinococcus multilocularis

Status: 2. Approximately 80 cases annually worldwide
Dist: NEA; PAL
Hab: Alveolar hydatid cyst in liver and other organs
Hosts: 2. Foxes *Vulpes*, *Alopex* spp. and other Carnivora
1. Rodents, especially *Microtus* spp.
Trans: Ingestion of egg by contamination
Zoo stat: 1
Coombs and Crompton 1991; Eckert 1998

Echinococcus oligarthrus

Status: 1. Very rare: first authenticated human case in 1989
Dist: NEO—Brazil, Suriname, Venezuela
Hab: Cyst intra-orbital or cardiac
Hosts: 2. Felidae: *Felis* spp.
1. Rodents: *Dasyprocta*, *Cuniculus paca*, *Proechimys guyannensis*
Trans: Ingestion of egg by contamination
Zoo stat: 1
Coombs and Crompton 1991; Eckert 1998; Basset *et al.* 1998

Echinococcus vogeli

Status: 1. Thirty-six human cases on record
Dist: NEO—Colombia, Costa Rica, Ecuador, Panama,Suriname, Venezuela
Hab: Cysts in liver etc.
Hosts: 2. Bush dog *Spatheotis venatus*
1. Rodents: *Cuniculus*, *Dasyprocta* spp.
Trans: Ingestion of egg by contamination
Zoo stat: 1
Beaver *et al.* 1984; Coombs and Crompton 1991; Eckert 1998; Basset *et al.* 1998

Multiceps brauni

Syn: *Taenia brauni*
Status: 1. Beaver *et al.* (1984) mentions 53 cases
Dist: AETH—Uganda etc.
Hab: Coenurus in eye and orbit
Hosts: 2. Dog and other Carnivora
1. Various rodents
Trans: Ingestion of egg by contamination
Zoo stat: 1
Beaver *et al.* 1984; Coombs and Crompton 1991

Multiceps glomeratus

Syn: *Taenia glomeratus*
Status: 1. One (possibly three) human cases on record
Dist: AETH—Nigeria
Hab: Cyst in muscles
Hosts: 2.
1. Rodents
Trans: Ingestion of egg by contamination
Zoo stat: 1
Beaver *et al.* 1984; Coombs and Crompton 1991

Multiceps longihamatus

Syn: *Taenia longihamatus*
Status: 1. Four human cases on record
Dist: PAL—Japan
Hab: Adult in small intestine
Hosts: 2.
1. Lagomorpha?
Trans: Ingestion of coenurus in rabbit or hare
Zoo stat: 2
Beaver *et al.* 1984; Coombs and Crompton 1991

Multiceps multiceps

Syn: *Taenia multiceps*
Status: 1. Six cases mentioned by Beaver *et al.* (1984)
Dist: AETH; NEO; NEA; PAL
Hab: Coenurus larva in eye or brain
Hosts: 2. Carnivora: Canidae
1. Ungulates
Trans: Ingestion of egg by contamination
Zoo stat: 1
Beaver *et al.* 1984; Coombs and Crompton 1991; Lloyd 1998a

Multiceps serialis

Syn: *Taenia serialis*
Status: 1. Four human cases on record
Dist: Parasite cosmopolitan
Hab: Bladder-like cysts in breast
Hosts: 2. Dog and other Carnivora: wolf, *Hyaena*
1. Rabbit, hare, rodents
Trans: Ingestion of egg by contamination
Zoo stat: 1
Beaver *et al.* 1984; Coombs and Crompton 1991; Lloyd 1998a

Taenia asiatica

Many authors consider this to be a pig-adapted subspecies of *T. saginata*
Status: 2. Recently distinguished but 'widely distributed in Asian countries'
Dist: PAL—Taiwan, Korea; OR—Thailand etc.
Hab: Adult worm in small intestine
Hosts: 2. Adult only known in man
1. Metacestode in viscera of pig, cow
Trans: Ingestion of egg by contamination
Zoo stat: 4
Eom and Rim 1993; Andreassen 1998; ICZN 2001

Taenia crassiceps

Status: 1. Two human cases reported
Dist: NEA—Canada; PAL—France
Hab: Larval cyst in eye, or embedded in deep tissues
Hosts: 2. Carnivora: *Felis lynx, Vulpes vulpes*
1. Rodents
Risk: The French case was severely immunocompromised
Trans: Ingestion of egg by contamination
Zoo stat: 1
Beaver *et al.* 1984; Coombs and Crompton 1991; Francois *et al.* 1998

Taenia saginata

Syn: *Taeniarhynchus saginatus*
Status: 4. Prevalence considerably greater than *T. solium*
Dist: Cosmopolitan
Hab: Adult in small intestine
Hosts: 2. Man
1. Ungulates: many species including cattle, *Rangifer*; rare in wild African ungulates
Trans: Ingestion of cysticercus in beef
Zoo stat: 4
Beaver *et al.* 1984; Coombs and Crompton 1991; Harrison and Sewell 1991

The rarity of *T. saginata* in indigenous African animals argues in favour of a Palaearctic or Oriental origin (*cf. T. solium* and *T. asiatica*). On the other hand, Hoberg *et al.* (2000) give convincing evidence that *T. solium* transferred from a sylvatic carnivore/herbivore cycle, to a human/herbivore cycle, in Africa, and subsequently transferred to cattle, in Eurasia, when these were domesticated.

Taenia solium

Status: 4. Stool positivity frequently >2%
Dist: Cosmopolitan
Hab: Adult in small intestine
Cysticercus larva in brain and subcutaneous tissue
Hosts: 2. Man
1. Domestic pig
Trans: Ingestion of cysticercus in pork
Ingestion of egg by faeco–oral contamination (direct)
Zoo stat: 4
Beaver *et al.* 1984; Coombs and Crompton 1991

Wild pigs are apparently unimportant as hosts in Africa (Harrison and Sewell 1991), which suggests a non-African origin for *T. solium*. On the other hand, Hoberg *et al.* (2000) give convincing evidence that *T. solium* transferred from a sylvatic carnivore/ herbivore cycle to a human/herbivore cycle, in Africa, and subsequently transferred to pigs, in Eurasia, when these were domesticated.

Taenia taeniaeformis

Status: 1. Two human cases on record
Dist: NEO—Argentina; PAL—'Czechoslovakia'
Hab: Larval cyst in liver
Hosts: 2. Cat, dog, other Carnivora: *Mustela foina* 'weasel' (*sic*)
1. Rodents
Trans: Ingestion of egg by contamination
Zoo stat: 1
Beaver *et al.* 1984; Coombs and Crompton 1991

PART 4: NEMATODA

PANAGROLAIMIDAE

Halicephalobus gingivalis

Syn: *Micronema deletrix*
Status: 1. Very few human cases
Dist: NEA—USA
Hab: Brain, liver, heart
Hosts: Normally free-living
Trans: Introduced through trauma
Zoo stat: 1
Anderson *et al.* 1998

RHABDITIDAE

Pelodera strongyloides

Four species of *Pelodera* are associated with mammals; it is doubtful if the human infection was correctly identified
Status: 1. A single recorded human infection
Dist: PAL—Poland
Hab: Larval worm in skin
Hosts: Free-living; larvae are phoretic in rodents
Trans:
Zoo stat: 1
Beaver *et al.* 1984; Coombs and Crompton 1991; Anderson 1992

Rhabditis elongata

Status: 1. Two human cases recorded
Dist: PAL—Korea
Hab: Alimentary tract
Hosts: Free-living, facultative parasite
Trans:
Zoo stat: 2
Coombs and Crompton 1991

Rhabditis inermis

Status: 1. A record of 17 infected school-children
Dist: PAL—Japan
Hab: Alimentary tract
Hosts: Free-living, facultative parasite
Trans:
Zoo stat: 2
Coombs and Crompton 1991

STRONGYLOIDIDAE

Strongyloides fuelleborni fuelleborni

Status: 3. Endemic, generally at low levels in forested areas
Dist: AETH—Ethiopia, Malawi, Namibia, Rwanda, Zambia, Zimbabwe, Central African Republic, Cameroon, Congo
Hab: Adult female in small intestine
Hosts: 2. Primates; probably maintained in man in some areas
1. Facultative free-living cycle
Trans: Larva actively penetrates skin; possible transmammary transmission
Zoo stat: 3
Pampiglione and Ricciardi 1972; Coombs and Crompton 1991; Ashford *et al.* 1992; ICZN 2001

Strongyloides fuelleborni kellyi

Status: 3. Frequently abundant in its very restricted range
Dist: AUS—Papua New Guinea
Hab: Adult female in small intestine
Hosts: 2. Only known from man
1. Facultative free-living cycle
Trans: Larva actively penetrates skin; transmammary transmission strongly suspected, but depends on circumstantial evidence
Zoo stat: 4
Coombs and Crompton 1991; Ashford *et al.* 1992

One other 'Zoo stat 4' is restricted to the Australasian Region: *Brugia timori*.

Strongyloides papillosus

Status: 1.
Dist: Cosmopolitan
Hab: Abortive larvae in skin
Hosts: 2. Ungulates, lagomorphs
1. Facultative free-living cycle
Trans: Larva actively invades skin
Zoo stat: 1
Coombs and Crompton 1991

Strongyloides ransomi

Status: 1.
Dist: Cosmopolitan
Hab: Abortive larva in skin
Hosts: 2. Pig
1. Facultative free-living cycle
Trans: Larva actively invades skin
Zoo stat: 1
Coombs and Crompton 1991

Strongyloides stercoralis

Status: 5. Abundant in warm humid climates; prevalence frequently >30% in adults
Dist: Cosmopolitan
Hab: Adult female in mucosa of small intestine
Hosts: 2. Man is the only known maintenance host
1. Facultative free-living cycle
Trans: Larva actively penetrates skin
Zoo stat: 4
Beaver *et al.* 1984; Coombs and Crompton 1991

Strongyloides westeri

Status: 1.
Dist: Cosmopolitan
Hab: Abortive larva in skin
Hosts: 2. Horse
1. Facultative free-living cycle
Trans: Larva actively invades skin
Zoo stat: 1
Coombs and Crompton 1991

ANCYLOSTOMATIDAE

Ancylostoma braziliense

Status: 3. 'A common causative agent of creeping eruption in the southern USA'
Dist: Cosmopolitan, but uncertain; identification a problem
Hab: Abortive larva in skin; hypobiotic larvae in muscle
Hosts: Carnivora: Canidae, Felidae
Trans: Larva actively invades skin
Zoo stat: 1
Beaver *et al.* 1984; Coombs and Crompton 1991; ICZN 2001

Ancylostoma caninum

Status: 2. Adult recorded >200 times in man
Dist: Cosmopolitan—human infection known in Philippines, Australia
Hab: Abortive larva in skin; hypobiotic larva in muscle; sterile adult in intestine
Hosts: Carnivora: Canidae, Felidae
Trans: Larva actively invades skin
Zoo stat: 1
Beaver *et al.* 1984; Coombs and Crompton 1991; Smyth 1995; Prociv 1998

Ancylostoma ceylanicum

Status: 2. Seems well adapted to the human host in Irian Jaya, Indonesia
Dist: PAL and OR, but uncertain due to doubtful identification
Hab: Adult in small intestine
Hosts: Carnivora: Canidae, Felidae
Trans: Larva actively penetrates skin; ingestion of larva by contami nation
Zoo stat: 2
Coombs and Crompton 1991; Prociv 1998

Ancylostoma duodenale

Status: 5. Endemic and locally abundant, usually less so than *Necator americanus*
Dist: Cosmopolitan—in sub-tropics and warm temperate zones
Hab: Attached to mucosa of small intestine
Hosts: No non-human host confirmed
Trans: Larva actively penetrates skin; ingestion of larva by contamination; possibly transplacental and transmammary transmission
Zoo stat: 4
Beaver *et al.* 1984; Coombs and Crompton 1991

Schad and Nawalinski (1991) cite Cameron (1951): 'one hookworm species, *Ancylostoma duodenale*, was acquired when man domesticated the dog'. (Around 12000 years ago according to Beck (2000).)

The evidence suggests that this species is not of Aethiopian origin: in Africa it is largely confined to coastal areas, where it may well have been introduced from Asia or the Middle East (see Goldsmid 1991).

Hawdon and Johnston (1996) review the possible routes of entry of this species to the Americas, and suggest that 'storm-tossed fishermen' are not the only candidates for pre-Columbian introduction, but that the species may have been introduced by the Beringian route. They suggest hypobiosis and vertical transmission as mechanisms facilitating survival of the Beringian migration.

Ancylostoma malayanum

Status: 1. A single human case record
Dist: OR—India, South-east Asia
Hab: Alimentary tract
Hosts: Bear
Trans:
Zoo stat: 2
Beaver *et al.* 1984; Coombs and Crompton 1991; Prociv 1998

Cyclodontostomum purvisi

Status: 1. A single human case record
Dist: OR—Thailand
Hab: Alimentary tract
Hosts: Rodents
Trans: Ingestion of larva by contamination
Zoo stat: 2
Beaver *et al.* 1984; Coombs and Crompton 1991

Necator americanus

Status: 6. The predominant hookworm in the tropics
Dist: Cosmopolitan—introduced in NEO and NEA
Hab: Adult attached to mucosa of small intestine
Hosts: Man is the only known maintenance host
Trans: Larva actively penetrates skin
Zoo stat: 4
Beaver *et al.* 1984; Coombs and Crompton 1991; ICZN 2001

Hawdon and Johnston (1996) accept the suggestion that this species was introduced to the Americas with the African slave trade. Schad (1991) suggests that *N. americanus* may have evolved from *N. suillus*, a parasite of pigs in South America and Trinidad which can infect man experimentally, but he also accepts that the parasite originated in the Old World, and was carried to the Americas.

CHABERTIIDAE

Oesophagostomum aculeatum

Syn: *Oesophagostomum apiostomum pro parte*
Status: 1. A single human record
Dist: OR—Indonesia
Hab: Intestine
Hosts: Primates
Trans:
Zoo stat: 2
Coombs and Crompton 1991; Polderman and Blotkamp 1995; ICZN 2001

Oesophagostomum bifurcum

Syn: *Oesophagostomum apiostomum pro parte*
Oesophagostomum brumpti
Status: 3. Very local, but prevalence >20% in endemic area
Dist: AETH—Northern Togo and Ghana
Hab: Larvae on mucosa and adults in lumen of large intestine
Hosts: Primates
Trans: (Ingestion of third-stage larvae)
Zoo stat: 3
Coombs and Crompton 1991; Polderman and Blotkamp 1995

Oesophagostomum stephanostomum

Status: 1. Three records in man
Dist: NEO; AETH
Hab: Intestinal abscesses
Hosts: Primates
Trans: Ingestion of larva by contamination
Zoo stat: 2
Coombs and Crompton 1991; Polderman and Blotkamp 1995

Ternidens deminutus

Status: 3. Prevalence may reach >80%, but very localised
Dist: AETH—Widespread in eastern and southern Africa
Hab: Larva in mucosa of large intestine; adult in lumen of large intestine
Hosts: Primates
Trans: Unknown (neither skin-penetration nor ingestion is successful)
Zoo stat: 3
Beaver *et al.* 1984; Coombs and Crompton 1991

The wide distribution of this species in non-human primates, in Asia as well as Africa, suggests that the form in humans, restricted to Africa, may be a different species. Taxonomic evidence is lacking, however (Goldsmid 1991).

SYNGAMIDAE

Mammomonogamus laryngeus

Status: 1. Total of 87cases recorded to 1995
Dist: AETH; NEO; OR
Hab: Adult in respiratory tract
Hosts: Ruminants, especially cow; Felidae
Trans: Uncertain
Zoo stat: 2
Beaver *et al.* 1984; Coombs and Crompton 1991; Freitas *et al.* 1995

Mammomonogamus nasicola

Status: 1. A single old human case record
Dist: NEO—West Indies
Hab: 'Pair of worms coughed up'
Hosts: Ungulates: sheep, cattle, goat, deer
Trans: (Ingestion of third-stage larva)
Zoo stat: 1
Beaver *et al.* 1984; Coombs and Crompton 1991

TRICHOSTRONGYLIDAE

Haemonchus contortus

Status: 1. Four human cases recorded by Beaver *et al.* (1984)
Dist: NEO—Brazil; PAL—Iran; AUS—Australia
Hab: Intestinal mucosa
Hosts: Ungulates
Trans: (Ingestion of third-stage larva)
Zoo stat: 2
Beaver *et al.* 1984; Coombs and Crompton 1991

Marshallagia marshalli

Status: 1. First report in 1973
Dist: PAL—Iran
Hab: Intestinal mucosa
Hosts: Ungulates
Trans: (Ingestion of third-stage larva)
Zoo stat: 2
Coombs and Crompton 1991

Ostertagia ostertagi

Status: 1. A single human case record
Dist: PAL—Iran, Azerbaijan
Hab: Intestine
Hosts: Ungulates: cattle
Trans: (Ingestion of larvae in nodules in abomasum)
Zoo stat: 2
Beaver *et al.* 1984; Coombs and Crompton 1991

Teladorsagia circumcincta

Syn: *Ostertagia circumcincta*
Status: 1. A single human case record
Dist: PAL—Azerbaidjan
Hab: Intestine
Hosts: Ungulates: goat, sheep etc.
Trans: (Ingestion of third-stage larva)
Zoo stat: 2
Beaver *et al.* 1984; Coombs and Crompton 1991; Anderson 1992

Trichostrongylus affinis

Status: 1.
Dist: Similar to that of rabbit
Hab: Intestine
Hosts: Lagomorpha: rabbit
Trans: (Ingestion of third-stage larva)
Zoo stat: 2
Coombs and Crompton 1991; ICZN 2001

Trichostrongylus axei

Status: 3. 'A few' among 40% prevalence of *T. orientalis* in Japan; common in man and other animals in Iran
Dist: AETH—Mauritius; OR—Indonesia; PAL—Iran, Armenia, Russian Federation: Siberia, Japan; AUS—Australia; NEO—Caribbean
Hab: Intestine
Hosts: Ungulates
Trans: (Ingestion of third-stage larva)
Zoo stat: 2
Beaver *et al.* 1984; Coombs and Crompton 1991

Trichostrongylus brevis

Status: 1. 'A few' among 40% prevalence of *T. orientalis*
Dist: PAL—Japan
Hab: Alimentary tract
Hosts: Only known from man
Trans: (Ingestion of third-stage larva)
Zoo stat: 2
Beaver *et al.* 1984; Coombs and Crompton 1991

Trichostrongylus calcaratus

Status: 1.
Dist: PAL—Iran
Hab: Intestine
Hosts: Lagomorpha: rabbit; Ungulates: sheep etc.
Trans: (Ingestion of third-stage larva)
Zoo stat: 2
Coombs and Crompton 1991

Trichostrongylus capricola

Status: 3. Common in man and other animals in Iran
Dist: PAL—Italy, Iran
Hab: Intestine
Hosts: Ungulates: goat, sheep etc.
Trans: (Ingestion of third-stage larva)
Zoo stat: 2
Beaver *et al.* 1984; Coombs and Crompton 1991

Trichostrongylus colubriformis

Status: 4. Prevalence reaches 70% in west Asia and Egypt
Dist: PAL—Egypt, Iran, Iraq, Japan, Italy, Armenia; OR—India, Indo nesia; AUS—Australia; NEA—USA: Louisiana
Hab: Intestine
Hosts: Ungulates: many species
Trans: (Ingestion of third-stage larva)
Zoo stat: 3
Beaver *et al.* 1984; Coombs and Crompton 1991

Trichostrongylus instabilis

Status: 1.
Dist: PAL—Armenia, Russian Federation: Siberia
Hab: Intestine
Hosts:
Trans: (Ingestion of third-stage larva)
Zoo stat: 2
Beaver *et al.* 1984; not in Nolan 1998

Trichostrongylus lerouxi

Status: 1.
Dist: PAL—Iran
Hab: Intestine
Hosts: Only known from man; normal hosts unknown
Trans: (Ingestion of third-stage larva)
Zoo stat: 2
Coombs and Crompton 1991; not in Nolan 1998

Trichostrongylus orientalis

Status: 4. In Japan, majority of trichostrongyles in 40% prevalence were of this species; now very rare; the joint predominant species in Iran where prevalence is 67%; man is the natural host
Dist: PAL—China, Iran, Armenia, Japan, Taiwan, Korea
Hab: Intestine
Hosts: 'Man, occasionally in ... domestic ruminants'; 'considered to be predominantly a parasite of man'
Trans: (Ingestion of third-stage larva)
Zoo stat: 3
Beaver *et al.* 1984; Coombs and Crompton 1991; Nolan 1998; Yoshimura 1998

Trichostrongylus probolurus

Status: 3. Common in man and animals in Iran
Dist: PAL—Egypt, Iran, Armenia, Russian Federation: Siberia
Hab: Intestine
Hosts: Ungulates
Trans: (Ingestion of third-stage larva)
Zoo stat: 2
Beaver *et al.* 1984; Coombs and Crompton 1991

Trichostrongylus skrjabini

Status: 3. Common in man and animals in Iran
Dist: PAL—Iran, Armenia
Hab: Intestine
Hosts: Ungulates: sheep, deer
Trans: (Ingestion of third-stage larva)
Zoo stat: 2
Beaver *et al.* 1984; Coombs and Crompton 1991

Trichostrongylus vitrinus

Status: 3. Common in man and animals in Iran
Dist: PAL—Morocco, Egypt, Iran, Armenia, Russian Federation: Siberia; NEO—Chile
Hab: Intestine
Hosts: Ungulates
Trans: (Ingestion of third-stage larva)
Zoo stat: 2
Beaver *et al.* 1984; Coombs and Crompton 1991

METASTRONGYLIDAE

Metastrongylus elongatus

Status: 1. Three human infections on record
Dist: Parasite cosmopolitan
Hab: Adult in respiratory tract
Hosts: 2. Pig, ruminants
1. Oligochaetes: *Lumbricus* sp.
Trans: Ingestion of larva
Zoo stat: 2
Beaver *et al.* 1984; Coombs and Crompton 1991

ANGIOSTRONGYLIDAE

Parastrongylus cantonensis

Syn: *Angiostrongylus cantonensis*
Status: 2. Hundreds of cases have now been reported
Dist: AETH; PAL; OR; AUS
Hab: Abortive larva migrates to brain
Hosts: 2. Rodents: *Rattus* spp.
1P. Crabs, frogs, shrimps
1. Mollusca: *Achatina*, *Pila*, *Veronicella* spp.
Trans: Ingestion of larva in paratenic host, or by contamination
Zoo stat: 1
Beaver *et al.* 1984; Coombs and Crompton 1991

Parastrongylus costaricensis

Syn: *Angiostrongylus costaricensis*
Status: 2. Scores of cases reported in Costa Rica; 116 in ten years in one hospital; 15 in other countries; incidence in Costa Rica estimated at 2000 per 100 000 per year
Dist: NEO—Costa Rica, Nicaragua, Panama, Peru
Hab: Adult in anterior mesenteric artery
Hosts: 2. Rodents: mainly *Sigmodon hispidus* but also many other species
1. Slug: *Vaginulus* sp.
Trans: Ingestion of larva by contamination
Zoo stat: 1
Beaver *et al.* 1984; Coombs and Crompton 1991; Cross 1998b

OXYURIDAE

Enterobius gregorii

Status: 6. Apparently more abundant than *E. vermicularis,* but rarely looked for
Dist: (Cosmopolitan)
Hab: Large intestine
Hosts: Man only
Trans: Ingestion of egg
Zoo stat: 4
Chittenden and Ashford 1987; Coombs and Crompton 1991

Hasegawa *et al*. (1998) suggest that *E. gregorii* is a juvenile form of *E. vermicularis*.

Enterobius vermicularis

Status: 6. Prevalence may reach 100% in children in temperate areas
Dist: Cosmopolitan
Hab: Large intestine
Hosts: Man only
Trans: Ingestion or inhalation of egg
Zoo stat: 4
Beaver *et al.* 1984; Coombs and Crompton 1991

E. gregorii and *E. vermicularis* are sister-species; the closest relative is *E. anthropopitheci* of chimpanzees *Pan satyrus* and *P. paniscus* (Hugot 1993).

Syphacea obvelata

Status: 1. A single human case on record
Dist: OR—Philippines
Hab: Intestine
Hosts: Rodents: mouse etc.
Trans: Ingestion of egg
Zoo stat: 2
Beaver *et al.* 1984; Coombs and Crompton 1991

ANISAKIDAE

Anisakis physeteris

('*Anisakis* type II larva')
Status: 1.
Dist:
Hab: Larva in mucosa of stomach and small intestine
Hosts: 2. Sperm whale *Physeter catodon*
1P. Marine fish and squid
1. Euphausiid crustacea
Trans: Ingestion of larva in fish or squid
Zoo stat: 1
Delyamure 1968; Yoshimura 1998

Anisakis simplex

('*Anisakis* type 1 larva')
Status: 2. Eleven cases recorded in Europe to 1966; >500 known in Japan, with around 1500 cases of anisakiasis annually in Japan
Dist: PAL—Japan, Baltic countries, The Netherlands; NEA—USA
Hab: Larva in mucosa of stomach and small intestine
Hosts: 2. Piscivorous marine mammals, predominantly Cetacea
1P. Marine fish and squid
1. Euphausiid crustacea
Trans: Ingestion of larva in fish or squid
Zoo stat: 1
Delyamure 1968; Beaver *et al.* 1984; Coombs and Crompton 1991; Yoshimura 1998

Contracaecum osculatum

Status: 1. One human case on record
Dist: PAL—Germany, Spain
Hab: Larva in mucosa of stomach and small intestine
Hosts: 2. Piscivorous marine mammals, predominantly Pinnipedia
1P. Marine fish
1. Euphausiid crustacea
Trans: Ingestion of larva in fish
Zoo stat: 1
Delyamure 1968; Beaver *et al.* 1984; Coombs and Crompton 1991

Hysterothylacium aduncum

Syn: *Contracaecum aduncum*
Status: 1.
Dist: NEO—Chile
Hab: Larva in mucosa of stomach
Hosts: 2. Marine fish
1P. Marine fish
1. Marine copepods and annelids
Trans:
Zoo stat: 1
Muller 2002

Phocanema decipiens

Syn: *Terranova decipiens*
Pseudoterranova decipiens
('*Terranova* type A larva'),
Status: 1. More than 23 cases recorded
Dist: NEA—Canada, USA; PAL—Japan, Korea
Hab: Larva in mucosa of stomach
Hosts: 3. Pinnipedia
2P. Marine fish, especially cod and halibut in Japan, rockfish, salmon and red snapper in the USA, and herring in The Netherlands; cephalopods
2. Marine invertebrates
1. Copepod crustacea
Trans: Ingestion of larva in fish or squid
Zoo stat: 1
Beaver *et al.* 1984; Coombs and Crompton 1991; Anderson 1992; Cheng 1998; Yoshimura 1998

Cheng (1998) states that human infection with any species of *Pseudoterranova* remains to be confirmed.

ASCARIDIDAE

Ascaris lumbricoides

Status: 6. Holo-endemic (prevalence >70%) in many in rural tropical communities
Dist: Cosmopolitan
Hab: Lumen of small intestine, following migration of larva
Hosts: Only maintained in man
Trans: Ingestion of egg following development in soil
Zoo stat: 4
Beaver *et al.* 1984; Coombs and Crompton 1991; ICZN 2001

Ascaris suum

Status: 1. Uncertain: sporadic cases at least
Dist: Cosmopolitan
Hab: Lumen of small intestine, following migration of larva
Hosts: Only maintained in pig
Trans: Ingestion of egg following development in soil
Zoo stat: 2
Beaver *et al.* 1984; Coombs and Crompton 1991

Anderson (1995) and Anderson and Jaenike (1997) confirm that, although no marker is known to distinguish individual specimens of *A. suum* from *A. lumbricoides*, sympatric populations are identifiable by gene frequencies and are reproductively isolated. Occasional infections in man in North America are *A. suum*.

Baylisascaris procyonis

Status: 1. First human infection recorded in 1984; sporadic cases since
Dist: NEA—USA
Hab: Migrating larvae in many tissues
Hosts: 2. Carnivora: raccoon *Procyon lotor*, with >40% prevalence in New York
1 (P?) Rodents may be paratenic or intermediate hosts
Trans: Ingestion of egg following development in soil
Zoo stat: 1
Coombs and Crompton 1991; Smyth 1995; Muller 2002

Lagochilascaris minor

Status: 1. Sixteen human cases on record
Dist: NEO—Brazil, Colombia, Costa Rica, Mexico, Suriname, Venezuela, Trinidad, Tobago
Hab: Larvae and adults in subcutaneous abscess
Hosts: Maintained experimentally using laboratory mouse as intermediate host and cat as final host
Trans: (Ingestion of egg)
Zoo stat: 1
Beaver *et al.* 1984; Coombs and Crompton 1991; Campos *et al.* 1995

Parascaris equorum

Status: 1.
Dist: Cosmopolitan in normal hosts
Hab: Migrating larva in many tissues
Hosts: Perissodactyla: horse
Trans: Ingestion of egg
Zoo stat: 1
Coombs and Crompton 1991

Toxocara canis

Status: 3. About 1900 cases reported by 1981; 2%–10% of people in Western countries are seropositive
Dist: Cosmopolitan in normal hosts
Hab: Migrating larva in many tissues
Hosts: Carnivora: dog
Trans: Ingestion of egg following development in soil
Zoo stat: 1
Beaver *et al.* 1984; Coombs and Crompton 1991; Lloyd 1998b

Toxocara cati

Status: 1. Twenty human cases reported
Dist: Cosmopolitan in normal host
Hab: Migrating larva in many tissues; immature adult in lumen of small intestine
Hosts: Carnivora: cat
Trans: Ingestion of egg following development in soil
Zoo stat: 1
Beaver *et al.* 1984; Coombs and Crompton 1991

DRACUNCULIDAE

Dracunculus medinensis

Status: 5 (now 2–3). Prevalence once reached 85% in some Indian villages; now rare except in Sudan
Dist: AETH; PAL
Hab: Larva migrates throughout body; adult in subcutaneous tissue
Hosts: 2. Man is the only confirmed maintenance host
1P. *Dracunculus insignis* can use frogs and tadpoles as paratenic hosts
1. Copepods: *Cyclops*, *Eucyclops*, *Macrocyclops* spp.
Trans: Ingestion of larva in *Cyclops* in drinking water
Zoo stat: 4
Beaver *et al.* 1984; Coombs and Crompton 1991; ICZN 2001

Dracunculus insignis of raccoons, mink and other carnivores in the USA and Canada may be the same species. This presents an interesting zoogeographical conundrum, unless the infection also occurs in Old World carnivores but is usually overlooked (as it was in America until specifically looked for). *Dracunculus lutrae* occurs in otters in Canada. *Dracunculus medinensis* is known from dogs but only in China. Two human cases in Japan and Korea, following consumption of fish, were presumably not caused by *D. medinensis* (Muller 1998).

GNATHOSTOMATIDAE

Gnathostoma binucleatum

Status: 2. Increasing prevalence, due to increased consumpton of farmed freshwater fish
Dist: NEO—Ecuador, Mexico and probably elsewhere
Hab: Larva in skin
Hosts: 3.
1P/2. Freshwater fish, various amphibia and reptiles
1. Freshwater copepods
Trans: Ingestion of larva in copepod or fish
Zoo stat: 1
McCarthy and Moore 2000; ICZN 2001

The finding that parasites in Mexico, initially identified as *G. hispidum*, are *G. binucleatum* suggests that the same applies to all New World infections.

Gnathostoma doloresi

Status: 1. First human cases—three confirmed, possibly eight in total—described in 1988
Dist: PAL—Japan
Hab: Larva in skin
Hosts: 3. Ungulates: pig
1P/2. Freshwater fish, various amphibia and reptiles
1. Freshwater copepods
Trans: Ingestion of larva in copepod or fish
Zoo stat: 1
Coombs and Crompton 1991

Gnathostoma hispidum

Status: 1. '...a human parasite on a few occasions'
Dist: PAL—China; OR—India
Hab: Larva in eye
Hosts: 3. Ungulates: pig
2. Freshwater fish, amphibia, mammals
1. Freshwater copepods
Trans: Ingestion of larva in host 1 or 2
Zoo stat: 1
Beaver *et al.* 1984; Coombs and Crompton 1991

Gnathostoma malaysiae

Status: 1. One case and one probable, in returnees to Japan from Myanmar
Dist: OR—Myanmar
Hab: Larvae in transient subcutaneous eruptions
Hosts: 3.
2.
1.
Trans: Ingestion of larva in freshwater shrimp
Zoo stat: 1
Nomura *et al.* 2000

Gnathostoma nipponicum

Status: 1. At least eight cases since 1988
Dist: PAL—Japan
Hab: Larvae in transient subcutaneous eruptions
Hosts: 3. Carnivora: weasel
2. Freshwater fish: loach
1. Freshwater copepod: *Cyclops* sp.
Trans: Ingestion of larva in host 2
Zoo stat: 1
Yoshimura 1998

Gnathostoma spinigerum

Status: 2
Dist: NEA—Mexico, USA: California; PAL—Japan; OR—Bangladesh, Cambodia, India, Indonesia, Laos, Malaysia, Myanmar, Philippines, Thailand, Vietnam
Hab: Larvae in transient subcutaneous swellings
Hosts: 3. Carnivora: cat, dog etc.
2. Freshwater crayfish, fish, amphibia, reptiles, birds
1. Freshwater copepods
Trans: Ingestion of larva in host 2
Zoo stat: 1
Coombs and Crompton 1991

PHYSALOPTERIDAE

Physaloptera caucasica

Syn: *Abbreviata caucasica*
Status: 1. Twelve cases listed by Beaver *et al.* (1984)
Dist: AETH—Congo, Namibia, Zambia, Zimbabwe; NEO—Brazil, Colombia, Panama; PAL—Israel; OR—India, Indonesia
Hab: Intestine
Hosts: 2. Primates
1. Unknown, presumably beetle or cockroach
Trans: (Ingestion of larva in intermediate or paratenic host)
Zoo stat: 2
Beaver *et al.* 1984; Coombs and Crompton 1991; Muller 2002

Physaloptera transfuga

Status: 1. Apparently just one human case on record
Dist: PAL—Russia
Hab: Larva subcutaneous in lip
Hosts: Cat, dog, amphibia, reptiles
Trans: ('Contact' with infective larva)
Zoo stat: 1
Coombs and Crompton 1991

RICTULARIIDAE

Rictularia sp.

Status: 1. A single human case record
Dist: NEA—USA: New York
Hab: Appendix
Hosts: 2. Rodents and bats
1. (Arthropod)
Trans: (Ingestion of larva in intermediate host)
Zoo stat: 2
Beaver *et al.* 1984; Coombs and Crompton 1991

THELAZIIDAE

Thelazia californiensis

Status: 1. Seven human cases on record
Dist: NEA—USA: California
Hab: Under eyelid
Hosts: 2. Ruminants, especially deer
1. Diptera
Trans: Introduction of larva into orbit by feeding fly
Zoo stat: 1
Beaver *et al.* 1984; Coombs and Crompton 1991; Muller 2002

Thelazia callipaeda

Status: 1. About 60 cases to 2002
Dist: PAL—China, Japan, Korea, Russia; OR—India, Indonesia, Myanmar, Thailand
Hab: Conjunctival sac
Hosts: 2. Dog, various mammals
1. Diptera.
Trans: Introduction of larva into orbit by feeding fly
Zoo stat: 1
Beaver *et al.* 1984; Coombs and Crompton 1991; Muller 2002

GONGYLONEMATIDAE

Gongylonema pulchrum

Status: 1. At least 40 human cases on record
Dist: PAL—Europe, Morocco, Spain, China; NEA—USA; OR—Sri Lanka; AUS—Australia, New Zealand
Hab: Larva and adult in mucosa and submucosa of buccal cavity
Hosts: 2. Pig, horse, camel
1. Coprophagous beetles
Trans: Ingestion of larva in intermediate host
Zoo stat: 2
Beaver *et al.* 1984; Coombs and Crompton 1991; Muller 2002

SPIROCERCIDAE

Spirocerca lupi

Status: 1. A single human case on record
Dist: PAL—Italy
Hab: 'Adults embedded in pockets of the terminal ileum'
Hosts: 2. Dog, other Canidae, Felidae
1. Coprophagous beetles, including *Scarabeus sacer*
Trans: Transplacental transfer of larva, mother having eaten intermediate host
Zoo stat: 2
Beaver *et al.* 1984; Coombs and Crompton 1991

ACUARIIDAE

Cheilospirura sp.

Status: 1. A single human case on record
Dist: OR—Philippines
Hab: Eye
Hosts: 2.
1.
Trans: Ingestion of larva in intermediate host
Zoo stat: 1
Beaver *et al.* 1984; Coombs and Crompton 1991

ONCHOCERCIDAE

Brugia beaveri

Status: 1. Six cases in N. America were probably this species
Dist: NEA—USA: Louisiana
Hab: Adult in lymph node
Hosts: 2. Carnivora: *Lynx*, *Mustela*, *Procyon* spp.
1. Presumably mosquitoes
Trans: Presumably active penetration of larva from feeding mosquito
Zoo stat: 1
Beaver *et al.* 1984; Coombs and Crompton 1991

Brugia sp. (*cf. B. ceylonensis*)

Status: 1. A single human case record
Dist: OR—Sri Lanka
Hab: Unfertilised adult in conjunctiva
Hosts: 2. Domestic dog
1.
Trans: Not described
Zoo stat: 1
Dissanaike *et al.* 2000

Brugia guyanensis

Status: 1. First human cases described in 1988
Dist: NEO—Peru
Hab: Adults in lymph nodes
Hosts: 2. Carnivora: coatimundi *Nasua nasua*
1. Unknown
Trans: Not described
Zoo stat: 1
Coombs and Crompton 1991

Brugia malayi

Status: 4. Periodic anthroponotic strain is holo-endemic, but subperiodic zoonotic form is less endemic
Dist: OR—India, Indonesia, Philippines, Thailand, Vietnam etc.
Hab: Microfilariae in blood; adult in lymphatics
Hosts: 2. Primates—especially *Presbytis* sp., cat, pangolin *Manis javanica*
1. Diptera: Culicidae: *Aedes*, *Anopheles*, *Mansonia* spp.
Trans: Active penetration of larva from feeding mosquito
Zoo stat: 3
Beaver *et al.* 1984; Coombs and Crompton 1991

Brugia pahangi

Status: 1. Only identifiable biochemically so poorly known; eight of nine *B. malayi* infections were mixed with *B. pahangi*
Dist: OR—Human infections only identified in Borneo, Indonesia
Hab: Microfilariae in blood; adult worms in lymphatics
Hosts: 2. Carnivora
1. Diptera: Culicidae
Trans: Active penetration of larva from feeding mosquito
Zoo stat: 2
Smyth 1995

The study cited by Smyth (1995) predates Coombs and Crompton (1991), but is not mentioned by them; Denham (1998) suggests that infection of man with this species remains to be confirmed.

Brugia timori

Status: 3. Prevalence reaches 30% in its restricted distribution
Dist: AUS—Lesser Sunda Islands
Hab: Microfilariae in blood; adult in lymphatics
Hosts: 2. Man is only known host
1. Diptera: Culicidae: *Anopheles* spp.
Trans: Active penetration of larva from feeding mosquito
Zoo stat: 4
Beaver *et al.* 1984; Coombs and Crompton 1991

The only other 'Zoo stat 4' restricted to the Australasian Region is *Strongyloides fuelleborni kellyi*.

Brugia sp. (*cf. B. tupaiae*)

Initially misidentified as *Brugia pahangi*
Status: 1. A single human case record
Dist: OR—Malaysia
Hab: Conjunctiva
Hosts: 2. Tree shrew *Tupaia* sp
1.
Trans: Not described
Zoo stat: 1
Dissanaike *et al.* 1974; Dissanaike *et al.* 2000

Dipetalonema sp. (*cf. D. arbuta/D. sprenti*)

Status: 1. Identity uncertain; three case records
Dist: NEA—USA: Oregon
Hab: Eye
Hosts: 2. Beaver *Castor canadensis* (*D. sprenti*); porcupine *Erethizon dorsatum* (*D. arbuta*)
1. Diptera: Culicidae: *Aedes* spp.
Trans: (Active penetration of larva from feeding vector)
Zoo stat: 1
Beaver *et al.* 1984; Coombs and Crompton 1991

Dirofilaria immitis

Syn: *Dirofilaria louisianensis*? *Dirofilaria magalhaesi*? *Dirofilaria spectans*
Status: 2. No more than 190 cases recorded to 1995
Dist: NEA; NEO; PAL—Japan; AUS
Hab: Immature adult usually in lung
Hosts: 2. Dog, other Carnivora
1. Diptera: Culicidae: *Aedes*, *Anopheles*, *Culex* spp.
Trans: Active penetration of infective larva from feeding vector
Zoo stat: 1
Beaver *et al.* 1984; Coombs and Crompton 1991; Pampiglione *et al.* 1995; S. Pampiglione, pers. comm.

Dirofilaria repens

Syn: *Dirofilaria conjunctivae pro parte*
Dirofilaria acutiuscula pro parte
Status: 2. More than 397 cases recorded to 1995
Dist: AETH; PAL; OR; in >30 countries
Hab: Immature or mature adult usually in subcutaneous or subconjunctival nodules; rarely in lung
Hosts: 2. Cat, dog, other Carnivora
1. Diptera: Culicidae: *Aedes*, *Anopheles*, *Culex* spp.
Trans: Active penetration of larva from feeding vector
Zoo stat: 1
Beaver *et al.* 1984; Coombs and Crompton 1991

Dirofilaria striata

Status: 1. A few human cases recorded
Dist: NEA—USA: North Carolina; NEO—Brazil, Venezuela
Hab: Adult in orbit
Hosts: 2. Carnivora: Felidae, Tayassuidae
1. Unknown; mosquitoes experimentally
Trans: (Active penetration of larva from feeding vector)
Zoo stat: 1
Orihel and Isbey 1990; Coombs and Crompton 1991

Dirofilaria tenuis

Syn: *Dirofilaria conjunctivae pro parte*
Status: 1. About 50 cases to 1995
Dist: NEA—USA: Florida, Mississippi and North Carolina
Hab: Immature adults mainly in subcutaneous tissue, also in orbit
Hosts: 2. Carnivora: raccoon *Procyon lotor*
1. Diptera: Culicidae: *Anopheles*, *Psorophora* spp.
Trans: (Active penetration of larva from feeding vector)
Zoo stat: 1
Beaver *et al.* 1984; Coombs and Crompton 1991; Pampiglione *et al.* 1995

Dirofilaria ursi

Status: 1. Ten cases provisionally identified with this species
Dist: NEA—Canada, USA
Hab: Immature adult in subcutaneous nodule
Hosts: 2. Carnivora: Ursidae: *Ursus* spp.
1. Diptera: Simuliidae: *Simulium* spp.
Trans: (Active penetration of larva from feeding vector)
Zoo stat: 1
Beaver *et al.* 1987

The identity of the worms is doubtful, and they could also be *D. subdermata* of porcupines.

Loa loa

Status: 4. Prevalence is often high and reaches 90% in some communities
Dist: AETH—from Senegal to Sudan
Hab: Microfilariae in blood; adult migrates in subcutaneous tissue
Hosts: 2. Possibly man only
1. Diptera: Tabanidae: *Chrysops* spp.
Trans: Active penetration of larva from feeding vector
Zoo stat: 3 or 4, depending on taxonomy
Beaver *et al.* 1984; Coombs and Crompton 1991; Boussinesq and Gardon 1997

Parasites in other primates are apparently different, and are known as *L. l. papionis* (Muller 2002)

Loaina sp.

Status: 1. A single case
Dist: NEO—Colombia
Hab: From the anterior chamber of the eye
Hosts: 2. Rabbits
1. Diptera: Culicidae
Trans: Active penetration of larva from feeding vector
Zoo stat: 1
Orihel and Eberhard 1998

Mansonella ozzardi

Status: 4. Prevalence may approach 90% in adults in isolated Amerindian communities
Dist: NEO—17 countries
Hab: Microfilariae in blood; adult in body cavities
Hosts: 2. Man is only known host
1. Diptera: *Culicoides*, *Simulium* spp.
Trans: Active penetration of larva from feeding vector
Zoo stat: 4—the only Neotropical 4
Beaver *et al.* 1984; Coombs and Crompton 1991

Mansonella perstans

Status: 5. Prevalence may reach 90%
Dist: AETH—widespread; NEO—east coast and Caribbean
Hab: Microfilariae in blood; adult in peritoneal cavity
Hosts: 2. Man, *Pan*
1. Diptera: Ceratopogonidae: *Culicoides* spp.
Trans: Active penetration of larva from feeding vector
Zoo stat: 4
Beaver *et al.* 1984; Coombs and Crompton 1991

Mansonella rodhaini

Status: 3. Fourteen of 440 people infected in one study
Dist: AETH—Gabon
Hab: Microfilariae in skin
Hosts: 2. Chimpanzees *Pan paniscus*, *Pan satyrus*
1. Unknown, probably *Culicoides* sp.
Trans: (Active penetration of larva from feeding vector)
Zoo stat: 2
Beaver *et al.* 1984; Coombs and Crompton 1991; Bain *et al.* 1995

Mansonella streptocerca

Status: 4. 44% prevalence in one study, but poorly known
Dist: AETH—Congo, Ghana, 'Zaire'
Hab: Microfilariae in skin; adult in subcutaneous tissue
Hosts: 2. Man, *Gorilla*, *Pan*
1. Diptera: Ceratopogonidae: *Culicoides* spp.
Trans: Active penetration of larva from feeding midge
Zoo stat: 4, possibly 3, but more information required
Beaver *et al.* 1984; Coombs and Crompton 1991

Meningonema peruzzii

Status: 1. Three cases recorded
Dist: AETH—Zimbabwe (?), Cameroon
Hab: Microfilariae and fourth-stage in cerebrospinal fluid
Hosts: 2. Primates: talapoin *Cercopithecus talapoin*
1. Unknown
Trans: (Active penetration of larva from feeding vector); microfilariae are in blood of reservoir host
Zoo stat: 1
Beaver *et al.* 1984; Coombs and Crompton 1991; Boussinesq *et al.* 1995

'Microfilaria' *bolivarensis*

Status: 1. Apparently a single report, but number of cases not given
Dist: NEO—Venezuela
Hab: Microfilariae in blood; adult unknown
Hosts: 2. Only known from man
1. Unknown
Trans: (Active penetration of larva from feeding vector)
Zoo stat: 2
Beaver *et al.* 1984; Coombs and Crompton 1991

'Microfilaria' *semiclarum*

Syn: *Mansonella semiclarum*
Status: 3. Restricted range, but reaches 50% locally
Dist: AETH—'Zaire'
Hab: Microfilariae in blood; adult unknown
Hosts: 2. Known only from man
1. Unknown
Trans: (Active penetration of larva from feeding vector)
Zoo stat: 2, but microfilariae differ from all others known—could even be 4
Beaver *et al.* 1984; Coombs and Crompton 1991

Onchocerca sp. (*cf. O. gutturosa*)

Status: 1. Eight cases reported in areas where *O. volvulus* is absent
Dist: NEA—Canada, USA; PAL—Albania, Japan, Russia, Switzerland
Hab: Adult in subcutaneous or ocular nodules
Hosts: 2. Cattle?
1. Diptera: Simuliidae: *Simulium* spp.?
Trans: Active penetration of larva from feeding blackfly
Zoo stat: 1
Orihel and Eberhard 1998; Pampiglione *et al.* 2001

The worms in all eight cases were similar to each other, and resembled *O. gutturosa*, a common parasite of domestic cattle.

Onchocerca volvulus

Status: 5. Holo-endemic in many communities
Dist: AETH—28 countries; NEO—Brazil, Colombia, Ecuador, Guatemala, Mexico, Venezuela
Hab: Microfilariae in skin; adult in subcutaneous nodules
Hosts: 2. Man, *Ateles, Gorilla*
1. Diptera: Simuliidae: *Simulium* spp.
Trans: Active penetration of larva from feeding blackfly
Zoo stat: 4
Beaver *et al.* 1984; Coombs and Crompton 1991

Setaria equina

Status: 1.
Dist: OR—India
Hab: Unknown; in natural hosts, adult is in peritoneal cavity
Hosts: 2. Camel, cattle, horse
1. *Aedes* spp.
Trans: (Active penetration of larva from feeding vector)
Zoo stat: 2
Coombs and Crompton 1991

Wuchereria bancrofti

Includes *Wuchereria lewisi*
Status: 5. Holo-endemic in widespread areas; prevalence frequently >50%
Dist: PAL; OR; AUS; NEO
Hab: Microfilariae in blood; adults in lymphatics
Hosts: 2. Man only
1. Diptera: Culicidae: *Aedes, Anopheles, Culex* spp.
Trans: Active penetration of larva from feeding mosquito
Zoo stat: 4
Beaver *et al.* 1984; Coombs and Crompton 1991

Failure of subsequent authors to confirm the existence of a distinctive species of *Wuchereria* in Brazil casts doubt on the validity of *W. lewisi.*

Closest known relative is *W. kalimantani* of *Presbytis cristatus* in Kalimantan.

DIOCTOPHYMATIDAE

Dioctophyme renale

Syn: *Dioctophyma renale*
Status: 1. At least 13 cases to 1984
Dist: NEA—Canada, USA; PAL—Iran, 'USSR'; OR—South-east Asia
Hab: Adult in kidney, but only one kidney affected in main hosts.
Hosts: 2. Carnivora, mainly mustelids and canids; main host in America is mink *Mustela vison*
1P. Freshwater fish: *Ameiurus*; frogs
1. Aquatic oligochaetes: *Cambarincola*, *Lumbriculus* spp.
Trans: Ingestion of larva in fish or frog flesh
Zoo stat: 2
Beaver *et al.* 1984; Coombs and Crompton 1991; Anderson 1992; ICZN 2001

Eustrongylides sp.

Status: 1. Three cases to 1984
Dist: NEA—USA
Hab: Larvae migrating through intestinal wall
Hosts: 3. Piscivorous birds such as *Ardea* spp.
2. Freshwater fish
1. Aquatic oligochaetes
Trans: Ingestion of larva in fish or oyster flesh
Zoo stat: 1
Beaver *et al.* 1984; Coombs and Crompton 1991

TRICHINELLIDAE

Trichinella britovi

Status: 1.
Dist: PAL—France, Italy; AETH—South Africa
Hab: Adult in intestine; larva encysted in striated muscle
Hosts: Fox *Vulpes* spp. and jackal *Canis* spp. are main hosts; also horse, pig
Trans: Ingestion of larva in flesh
Zoo stat: 1
Pozio *et al.* 1992

Trichinella murrelli

Status: 1. The cause of sylvatic trichinellosis in North America
Dist: NEA
Hab: Adult in intestine; larva encysted in striated muscle
Hosts:
Trans: Ingestion of larva in flesh
Zoo stat: 1
Pozio and La Rosa 2000

T. murrelli is very close to *T. britovi*; one of the 'distinguishing features' is its Nearctic distribution, so it seems likely to be synonymised in due course

Trichinella nativa

Status: 1. Rare, small, single-source outbreaks
Dist: NEA—Greenland, USA: Alaska; PAL—Arctic and sub-Arctic regions
Hab: Adult in intestine; larva encysted in striated muscle
Hosts: Carnivora: wolf, fox, raccoon, bears; Pinnipedes: walrus *Odobaenus rormarus*, seals
Trans: Ingestion of larva in flesh
Zoo stat: 1
Coombs and Crompton 1991

Trichinella nelsoni

Status: 1. Sporadic outbreaks; probably 100 or so cases reported to 1991
Dist: AETH—tropical Africa only
Hab: Adult in intestine; larva encysted in striated muscle
Hosts: Main host, Carnivora: hyaena *Crocuta crocuta*; also Ungulata: bushpig, warthog; various large carnivores and wild but not domestic Suidae
Trans: Ingestion of larva in flesh
Zoo stat: 1
Campbell 1991; Coombs and Crompton 1991; Pozio *et al.* 1992

Trichinella pseudospiralis

Status: 1. One isolated case, associated with Tasmania, and a small outbreak in Thailand
Dist: Parasite cosmopolitan
Hab: Adult presumably in intestine; larva unencysted, in striated muscle
Hosts: Mainly birds, but Thai cases associated with pig
Trans: Ingestion of larva in pork
Zoo stat: 1
Dennett *et al.* 1998; Jongwutiwes *et al.* 1998

The Tasmanian (?) case was also infected with an unidentified Muspiceoid nematode.

Trichinella spiralis

Status: 3. About 100 cases annually in USA; usually occurs in short-lived outbreaks
Dist: Cosmopolitan (except AUS?)
Hab: Adult in intestine, larva encysted in striated muscle
Hosts: Cat, dog, pig, bear, mouse, raccoon etc.
Trans: Ingestion of larva in flesh, especially pork
Zoo stat: 1
Beaver *et al.* 1984; Coombs and Crompton 1991

TRICHURIDAE

Anatrichosoma cutaneum

Status: 1. Two cases to 1984
Dist: PAL—Japan; OR—Malaysia, Vietnam
Hab: Adult in burrows in skin
Hosts: 1. Primates: *Macaca* sp.
Trans:
Zoo stat: 2
Beaver *et al.* 1984; Coombs and Crompton 1991; Muller 2002

Calodium hepaticum

Syn: *Capillaria hepatica*
Status: 1. Twenty-eight cases to 1984
Dist: AETH—South Africa; NEA—USA; NEO—Brazil, Mexico; OR—India; PAL—Czechoslovakia, Italy, Korea, Turkey
Hab: Adult in liver
Hosts: Numerous rodents, occasional in other mammals
Trans: Eggs pass through predator, without developing, to reach environment, then embryonate and infect by contamination
Zoo stat: 1
Beaver *et al.* 1984; Coombs and Crompton 1991; Anderson 1992

Eucoleus aerophilus

Syn: *Capillaria aerophila*
Status: 1. Eighteen cases to 1984
Dist: PAL—Morocco, Iran, 'USSR'; NEA; AUS
Hab: Adult in respiratory tract
Hosts: 2. Fox is apparently the main host; cat, dog, and other carnivores are also infected
1. Earthworms
Trans:
Zoo stat: 2
Beaver *et al.* 1984; Coombs and Crompton 1991; Cross 1998a

Paracapillaria philippinensis

Syn: *Aonchotheca philippinensis*
Calodium philippinensis
Capillaria philippinensis
Status: 3. Hundreds of cases in outbreak in Philippines
Dist: PAL—Iran, Japan, Korea; OR—Philippines, Thailand
Hab: Adult in small intestine
Hosts: 2. Unknown, piscivorous birds are strongly suspected
1. Brackish and freshwater fish: *Hypselotris* sp.
Trans: Ingestion of larva in fish
Zoo stat: 2
Coombs and Crompton 1991; Anderson 1992; Cross 1998a; Moravec 2001

Trichuris suis

Status: 1. 'Reported rarely from human intestine'
Dist: Cosmopolitan
Hab: Large intestine
Hosts: Pig
Trans: Ingestion of egg by contamination
Zoo stat: 2
Beaver *et al.* 1984; Coombs and Crompton 1991

Ooi *et al.* (1993) discovered a character to distinguish *T. suis* from *T. trichiura*

Trichuris trichiura

Status: 6. Holo-endemic in many tropical communities
Dist: Cosmopolitan
Hab: Large intestine
Hosts: Primates
Trans: Ingestion of egg by faeco–oral contamination (delayed)
Zoo stat: 3
Beaver *et al.* 1984; Coombs and Crompton 1991.

Ooi *et al.* (1993) were unable to distinguish conclusively between *T. trichiura* of human and monkey origin.

Trichuris vulpis

Status: 1. Rare as a human infection, and identification uncertain
Dist: NEA—USA; OR—Andaman Island; PAL—Europe (parasite cosmopolitan)
Hab: Large intestine
Hosts: Canidae
Trans: Ingestion of egg by faeco–oral contamination (delayed)
Zoo stat: 2
Coombs and Crompton 1991; Singh *et al.* 1993

Despite the finding by Yoshikawa *et al.* (1989), of eggs resembling those of *T. vulpis,* produced occasionally by *T. trichiura*, we accept the probability that *T. vulpis* does occasionally infect humans.

MUSPICEOIDEA

Genus and species unidentified

Status: 1. Two cases reported
Dist: AUS—both cases associated with Tasmania
Hab: Muscles
Hosts:
Trans:
Zoo stat: 1
Dennett *et al.* 1998

One of the cases was also apparently infected with *Trichinella pseudospiralis*: a single person with two previously unrecorded human parasites!

Part 5: Acanthocephala

ACANTHOCEPHALA

Acanthocephalus bufonis

Syn: *Acanthocephalus sinensis*
Pseudoacanthocephalus bufonis
Status: 1. A single human infection on record, in 1954
Dist: OR—Java
Hab: Small intestine
Hosts: 2. Amphibia *Bufo* spp.
1.
Trans: Ingestion of larva in intermediate or paratenic host
Zoo stat: 1
Schmidt 1971; Coombs and Crompton 1991

Acanthocephalus rauschi

Status: 1. A single human infection on record to 1984
Dist: NEA—USA: Alaska
Hab: Peritoneal cavity
Hosts: 2. Unknown, perhaps marine fish
1P. Unknown
1. Unknown
Trans: Ingestion of larva in intermediate or paratenic host
Zoo stat: 1
Beaver *et al.* 1984; Coombs and Crompton 1991

Bolbosoma sp.

Status: 1. Two records to 1984
Dist: PAL—Japan
Hab: Jejunum
Hosts: 2. Unknown, probably Cetacea
1P. Unknown, probably fish
1. Unknown, probably crustacean
Trans: Ingestion of juvenile in fish
Zoo stat: 2
Beaver *et al.* 1984; Coombs and Crompton 1991

Corynosoma strumosum

Status: 1. A single human infection, 1971, but 'probably fairly common ...'
Dist: NEA: Alaska
Hab: Immature adult in intestine
Hosts: 2. Seals
1P. Many species of fish
1.
Trans: Ingestion of larva in intermediate or paratenic host
Zoo stat: 1.
Schmidt 1971; Coombs and Crompton 1991

Macracanthorhynchus hirudinaceus

Status: 1. Five cases reported to 1984
Dist: AETH—Madagascar; PAL—China, 'USSR'; OR—Thailand
Hab: Small intestine
Hosts: 2. Pig
1. Various insects
Trans: Ingestion of larva in intermediate host
Zoo stat: 2
Beaver *et al.* 1984; Coombs and Crompton 1991

Macracanthorhynchus ingens

Status: 1.
Dist: NEA—USA: Texas
Hab: Small intestine
Hosts: 2. Carnivora including raccoon *Procyon lotor*, skunk *Mephitis mesomelas*
1P. Various frogs and snakes
1. Various terrestrial arthropods
Trans: Ingestion of larva in intermediate host
Zoo stat: 2
Coombs and Crompton 1991

Moniliformis moniliformis

Syn: *Moniliformis dubius*
Status: 1. Solitary human infections in nine countries
Dist: Parasite cosmopolitan
Hab: Small intestine
Hosts: 2. Rodents, especially *Rattus* spp.
1P. Amphibia and reptiles: *Bufo*, *Ctenosaura* spp.
1. Insects, including *Periplaneta*, *Geotrupes* and *Blaps*
Trans: Ingestion of cystacanth larva in intermediate host
Zoo stat: 2
Beaver *et al.* 1984; Coombs and Crompton 1991

Part 6: Arthropoda

CRUSTACEA: PENTASTOMIDA

Armillifer agkistrodontis

Status: 1. A single case report
Dist: PAL—China
Hab:
Hosts: 2. *Agkistrodon acutus*
1.
Trans: Ingestion of snake bile or blood, or of food contaminated with eggs from snake faeces or secretions
Zoo stat: 1
Zhang *et al.* 1996

Armillifer armillatus

Status: 3. Prevalence reaches 20% in eastern 'Zaire'
Dist: AETH
Hab: Nymph in liver, spleen, lungs; larva in eye
Hosts: 2. Pythons *Python* spp. and vipers *Bitis* spp.
1. Many species of mammal
Trans: Ingestion of snake, or of food contaminated with eggs from snake faeces or secretions
Zoo stat: 1
Beaver *et al.* 1984; Riley 1986; Lane and Crosskey 1993

Armillifer grandis

Status: 1. Nine cases on record
Dist: AETH—Congo basin
Hab:
Hosts: 2. 'Shares some snake hosts with *A. armillatus*'
1.
Trans: Ingestion of snake, or of food contaminated with eggs from snake faeces or secretions
Zoo stat: 1
Beaver *et al.* 1984; Riley 1986; Lane and Crosskey 1993

Armillifer moniliformis

Status: 3. Locally common, with prevalence >40% in Malaysian aboriginals
Dist: OR—Malaysia
Hab:
Hosts: 2. *Python* spp.
1.
Trans: Ingestion of snake, or of food contaminated with eggs from snake faeces or secretions
Zoo stat: 1
Beaver *et al.* 1984; Riley 1986; Lane and Crosskey 1993

Linguatula serrata

Status: 2. Rare but widespread
Dist: AETH; NEA; NEO; PAL
Hab: Larvae in tissue cysts, especially in mesenteric lymph nodes; adults in upper respiratory tract and sinuses
Hosts: 2. Carnivora: *Canis*, *Vulpes* spp.
1. Many herbivores
Trans: Ingestion of herbivores, or of eggs in contaminated food
Zoo stat: 2
Beaver *et al.* 1984; Riley 1986; Lane and Crosskey 1993

Pentastoma najae

Status: 1. A single human case
Dist: OR
Hab:
Hosts: 2. Snakes
Trans: Ingestion of snake, or of food contaminated with eggs from snake faeces or secretions
Zoo stat: 1
Riley 1986; Lane and Crosskey 1993

ACARI

Cheyletiella blakei

Status: 2.
Dist:
Hab: Feeds on epidermal keratin; human infection abortive
Hosts: Cat
Trans: Active transfer during close contact with pets
Zoo stat: 1
Beesley 1998

Cheyletiella furmani

Status: 2.
Dist:
Hab: Feeds on epidermal keratin; human infection abortive
Hosts: Rabbit
Trans: Active transfer during close contact with pets
Zoo stat: 1
Beesley 1998

Cheyletiella parasitivorax

Status: 2.
Dist: PAL
Hab: Feeds on epidermal keratin; human infection abortive
Hosts: Rabbit
Trans: Active transfer during close contact with pets
Zoo stat: 1
Beesley 1998

Cheyletiella yasguri

Status: 2
Dist: PAL
Hab: Feeds on epidermal keratin; human infection abortive
Hosts: Dog
Trans: Active transfer during close contact with pets
Zoo stat: 1
Beesley 1998

Demodex brevis

Status: 6. Presumably abundant in all populations, but rarely investigated
Dist: Cosmopolitan
Hab: Sebaceous glands
Hosts: Man is the only known host
Trans: Active transfer during prolonged personal contact
Zoo stat: 4
Beaver *et al.* 1984; Beesley 1998; ICZN 2001

Demodex folliculorum

Status: 6. Presumably abundant in all populations, but rarely investigated
Dist: Cosmopolitan
Hab: Hair follicles
Hosts: Man is the only known host
Trans: Active transfer during prolonged personal contact
Zoo stat: 4
Beaver *et al.* 1984; Beesley 1998

Sarcoptes bovis

Cause of 'Milker's itch'
Status: 2.
Dist: Cosmopolitan
Hab: Abortive superficial infection
Hosts: Cow
Trans: Active transfer during prolonged contact with cows
Zoo stat: 1
Beesley 1998; ICZN 2001

Sarcoptes equi

Cause of 'Cavalryman's itch'
Status: 2.
Dist: Cosmopolitan
Hab: Abortive superficial infection
Hosts: Horse
Trans: Active transfer during prolonged contact with horses
Zoo stat: 1
Beesley 1998

Sarcoptes scabiei

Status: 6. Common in most communities
Dist: Cosmopolitan
Hab: Burrows in skin
Hosts: Man is the only host but morphologically similar forms occur on many animals
Trans: Active transfer during prolonged personal contact
Zoo stat: 4
Beaver *et al.* 1984; Lane and Crosskey 1993; Beesley 1998

Trombiculid mites of the genera *Trombicula*, *Neotrombicula*, *Eutrombicula*, *Leptotrombidium* and *Ascoschoengastia* feed on man as larvae. Although they spend a few days feeding, they are perhaps better regarded, like **ticks** and the **dermanyssid mites**, as micropredators rather than as parasites for the present purposes.

INSECTA: ANOPLURA

Pediculus capitis

Head louse
Status: 6.
Dist: Cosmopolitan
Hab: Head hairs
Hosts: Known from man only
Trans: Active transfer during close contact
Zoo stat: 4
Schaefer 1978; Beaver *et al.* 1984; Lane and Crosskey 1993; ICZN 2001

Pediculus humanus

Body louse
Status: 6.
Dist: Cosmopolitan
Hab: Surface of skin and hairs
Hosts: Known from man only
Trans: Active transfer in clothing or during close contact
Zoo stat: 4
Schaefer 1978; Beaver *et al.* 1984; Lane and Crosskey 1993

Schaefer (1978) suggests that the two forms, *P. capitis* and *P. humanus*, should be regarded as separate species: although they are sympatric and only partially ecologically isolated, they rarely if ever hybridise. He cites Busvine as being of the same opinion.

Pthirus pubis

Pubic louse, crab louse
Status: 6.
Dist: Cosmopolitan
Hab: Pubic, peri-anal and coarse body hairs
Hosts: Known from man only
Trans: Active transfer in clothing or during close, often venereal, contact
Zoo stat: 4
Beaver *et al.* 1984; Lane and Crosskey 1993; ICZN 2001

The only other *Pthirus* species is *P. gorillae*, of the gorilla, of which only the female is known.

INSECTA: DIPTERA

Numerous species are recorded as accidental inhabitants of wounds, or of the intestine. Only those forms which invade living tissue are included here. Others mentioned by Beaver *et al.* (1984), which cause myiasis in necrotic tissue but do not invade healthy tissue, are listed as 'semiparasites' in the appendix.

DIPTERA: STRATIOMYIDAE

Hermetia illucens

Status: 1. Two case reports
Dist: AETH?; OR—Malaysia
Hab: Larva in cutaneous nodule or enteric
Hosts: 2. Adult free-living
1. Larva normally free-living on carrion
Trans: Larva presumably deposited on clothing
Zoo stat: 1?
Adler and Brancato 1995; Lee *et al.* 1995

DIPTERA: SARCOPHAGIDAE

Sarcophaga carnaria

Status: 2. 'A common cause of myiasis…'
Dist: PAL—Very widespread
Hab: Larvae attack wounds and body openings
Hosts: 2. Adult free-living
1. Larva normally free-living
Trans: Larva deposited on wound or in opening
Zoo stat: 2
James 1947; Beaver *et al.* 1984

Sarcophaga haemorrhoidalis

Syn: *Sarcophaga cruentata*
Status: 2.
Dist: AETH; AUS—(Introduced); NEA; PAL
Hab: Larvae mostly in wounds, migrating to healthy tissue; sometimes developing in intestine
Hosts: 2. Adult free-living
1. Larva usually free-living
Trans: Larva deposited on wound or ingested on contaminated food
Zoo stat: 2
James 1947; Beaver *et al.* 1984

Wohlfahrtia magnifica

Status: 3. 'A scourge in the steppes...'
Dist: PAL—Mainly in south-eastern Europe, western Asia and North Africa
Hab: Larvae move from superficial wounds and openings to invade healthy tissue
Hosts: 2. Adult free-living
1. Larva in domestic livestock
Trans: Female deposits larva on wound or opening
Zoo stat: 2
James 1947; Beaver *et al.* 1984

Wohlfahrtia vigil

Status: 2.
Dist: NEA—Especially mid-western USA, Minnesota, North Dakota, South Dakota, and Ontario, Canada
Hab: Larva in open lesions in skin
Hosts: 1. Adult free-living
2. Larva in young wild animals, especially mink
Trans: Female deposits larva on healthy skin
Risk: Predominantly in babies under five months
Zoo stat: 2
James 1947; Beaver *et al.* 1984

DIPTERA: CALLIPHORIDAE

Auchmeromyia senegalensis

Congo floor maggot
Syn: *Auchmeromyia luteola*
Status: 3. Strictly a micropredator rather than a parasite; 'will disappear with the advent of civilisation'
Dist: AETH—Mainly Equatorial forests
Hab: Larva lives on floor of houses and sucks blood
Hosts: 2. Adult free-living
1. Larva free-living in cracks in floor. 'The only ectoparasitic fly larva adapted to feeding on man', but also feeds on pig, warthog and aardvark
Trans: Eggs laid in houses; larvae actively seek sleeping host
Zoo stat: 3
James 1947; Beaver *et al.* 1984; Beesley 1998

Calliphora vicina

Bluebottle
Status: 2. Facultative parasite; uncommon as a parasite
Dist: Cosmopolitan in temperate regions
Hab: Usually free-living but may attack diseased tissue and invade surrounding tissue; occasional gastro-intestinal cases are probably pseudoparasitic
Hosts: 2. Adult free-living
1. Larva usually on carrion; sometimes on sheep
Trans: Larva enters through diseased tissue
Risk: Only attacks diseased tissue
Zoo stat: 2
James 1947; Beaver *et al.* 1984; ICZN 2001

Calliphora vomitoria

Status: 2. Facultative parasite
Dist: AETH; AUS; NEA; OR; PAL. Predominantly temperate regions
Hab: Usually free-living but may attack wounds and penetrate surrounding tissue; occasional gastrointestinal cases are probably pseudoparasitic
Hosts: 2. Adult free-living
1. Larva usually on carrion
Trans: Eggs deposited on necrotic tissue
Zoo stat: 2
James 1947; Beaver *et al.* 1984

Chrysomya bezziana

Old World screwworm
Status: 2. Frequently attacks man
Dist: AETH; OR
Hab: Larva burrows deeply into living tissues at wound site or orifice
Hosts: 2. Adult free-living
1. Larva in cattle and other mammals
Trans: Egg laid close to margin of a wound
Zoo stat: 2
James 1947; Beaver *et al.* 1984; Lane and Crosskey 1993

Chrysomya megacephala

Oriental latrine fly
Status: 2. Facultative parasite; 'known to cause myiasis in humans'
Dist: AETH—(Introduced); AUS; NEO—(Introduced); OR
Hab: Larva burrows deeply into living tissues at wound site or orifice
Hosts: 2. Adult free-living
1. Larva usually on faeces or carrion
Trans:
Zoo stat: 2
Lane and Crosskey 1993; Hall and Wall 1995

Cochliomyia hominivorax

Primary screwworm
Syn: *Callitroga americana*
Status: 3. Man is often attacked
Dist: NEA; NEO
Hab: Healthy tissue surrounding a clean wound.
Hosts: 2. Adult free-living
1. Larva in most domestic herbivores
Trans: Eggs laid close to a fresh clean wound
Zoo stat: 2
James 1947; Beaver *et al.* 1984; ICZN 2001

Cochliomyia macellaria

Secondary screwworm; New World screwworm
Status: 2. Facultative parasite; can cause myiasis in humans
Dist: NEA; NEO
Hab: Healthy tissue surrounding a clean wound
Hosts: 2. Adult free-living
1. Larva usually in carrion
Trans:
Risk: Immobile or debilitated people
Zoo stat: 2
Lane and Crosskey 1993; Hall and Wall 1995

Cordylobia anthropophaga

Tumbu fly
Status: 4. 'Common in man ...'
Dist: AETH—Widespread
Hab: Larva develops in cutaneous nodules
Hosts: 2. Adult free-living
1. Larva usually in dogs, also rats
Trans: Eggs laid on soil contaminated with urine or faeces, or on soiled clothes; larva actively penetrates healthy skin
Zoo stat: 2
James 1947; Beaver *et al.* 1984; Beesley 1998

Cordylobia rodhaini

Syn: *Stasisia rodhaini*
Status: 2. Man is an abnormal but occasional host
Dist: AETH—Equatorial countries, Ethiopia
Hab: Larva develops in cutaneous nodules
Hosts: 2. Adult free-living
1. Larva in antelopes and rodents
Trans: Larva actively penetrates healthy skin
Zoo stat: 2
James 1947; Beaver *et al.* 1984

Phaenicia sericata

Greenbottle, sheep fly
Syn: *Lucilia sericata*
Status: 2. Facultative parasite; American strain in wounds only
Dist: Cosmopolitan—mainly in temperate regions
Hab: Normally on carrion; in wounds, young larvae on surface, but older larvae may bore deeply into healthy tissue
Hosts: 2. Adult free-living
1. Larva usually on carrion but frequently on sheep
Trans: Eggs laid on diseased tissue
Zoo stat: 1
James 1947; Beaver *et al.* 1984

DIPTERA: OESTRIDAE: GASTEROPHILINAE

Gasterophilus haemorrhoidalis

Status: 1. Known cases are not common
Dist: Cosmopolitan
Hab: Abortive larva burrows in skin; migrates to intestine in normal hosts
Hosts: 2. Adult free-living
1. Larva: horse is normal host
Trans: Eggs laid close to mouth; larva burrows into skin
Zoo stat: 1
James 1947; Beaver *et al.* 1984; ICZN 2001

Gasterophilus intestinalis

Common horse botfly
Status: 1. 'A number of human cases are on record'
Dist: Cosmopolitan
Hab: Abortive larva burrows in superficial tissues; in normal hosts, larva burrows into mucosa
Hosts: 2. Adult free-living
1. Larva: horse is normal host
Trans: Eggs laid on hairs hatch when in contact with lips
Zoo stat: 1
James 1947; Beaver *et al.* 1984

Gasterophilus nigricornis

Status: 1. A single case recorded
Dist: PAL—China
Hab: Abortive larva burrows in skin
Hosts: 2. Adult free-living
1.
Trans: Egg laid on skin; larva burrows into tissues
Zoo stat: 1
Ning 2001

DIPTERA: OESTRIDAE: CUTEREBRINAE

Alouattamyia baeri

Status: 1. 'Can parasitise man'
Dist: NEO
Hab: Larva in lung; 'can cause pulmonary myiasis'
Hosts: 2. Adult free-living
1. Larva in howler monkey
Trans:
Zoo stat: 2
Hall and Wall 1995

Cuterebra sp.

Status: 1. Some 30 cases to 1999
Dist: NEA
Hab: Abortive larva in skin
Hosts: 2. Adult free-living
1. Rodents, lagomorphs
Trans: Egg laid on skin or mucosa; larva burrows into subcutaneous tissue
Zoo stat: 1
Keith 1999

Dermatobia hominis

Human botfly
Syn: *Dermatobia cyaniventris*
Status: 3. A common pest; increasingly a travel problem
Dist: NEO—all countries
Hab: Larva in subcutaneous lesion
Hosts: 2. Adult free-living
1. Larva in various mammals; can be a serious pest of cattle, but rarely horses
Trans: Eggs laid on phoretic mosquito; larva penetrates actively
Zoo stat: 2
James 1947; Beaver *et al.* 1984; Hall and Wall 1995

DIPTERA: OESTRIDAE: HYPODERMATINAE

Hypoderma bovis

Warble fly
Status: 2. 'Numerous records from N. America and Europe'
Dist: NEA; PAL
Hab: Larva migrates under skin to form warble on back; it can migrate into the eye in man
Hosts: 2. Adult free-living
1. Larva in cattle
Trans: Eggs attached to hairs, frequently near eyes; larva penetrates skin
Zoo stat: 2
James 1947; Beaver *et al.* 1984; ICZN 2001

Hypoderma diana

Deer warble fly; deer botfly
Status: 2. 'Numerous records'
Dist: PAL—Central and southern Europe
Hab: As *H. bovis*
Hosts: 2. Adult free-living
1. Larva usually in red or roe deer
Trans: Eggs attached to hairs, frequently near eyes; larva penetrates skin
Zoo stat: 2
James 1947

Hypoderma lineatum

Warble fly
Status: 2. 'Numerous records'
Dist: AETH—(Introduced); NEA; NEO—(Introduced); PAL
Hab: Larva migrates under skin, lodges in oesophagus before reaching back
Hosts: 2. Adult free-living
1. Larva in cattle
Trans: Eggs attached to hairs, frequently near eyes; larva penetrates skin
Zoo stat: 2
James 1947; Beaver *et al.* 1984

Hypoderma tarandi

Reindeer warble fly
Status: 2. 'Increasingly reported since 1982'
Dist: PAL—Norway; NEA—Canada, USA: Alaska
Hab: Larva in eye
Hosts: 2. Adult free-living
1. Larva usually in reindeer
Trans: As *H. bovis*
Zoo stat: 2
Hall and Wall 1995

DIPTERA: OESTRIDAE: OESTRINAE

Oestrus ovis

Sheep botfly
Status: 3. Man is often infested
Dist: Cosmopolitan
Hab: Abortive larva, either in conjunctiva or in nose, pharynx and larynx
Hosts: 2. Adult free-living
1. Larva in sheep and goats
Trans: Larva deposited around nose or lips
Zoo stat: 1
James 1947; Pampiglione 1958; Beaver *et al.* 1984; ICZN 2001

Pharyngomia picta

Red deer throat botfly
Status: 1. 'Has been reported in man'
Dist:
Hab: Abortive larva in conjunctiva
Hosts: 2. Adult free-living
1.
Trans:
Zoo stat: 1
Hall and Wall 1995

Rhinoestrus purpureus

Russian gadfly
Status: 2. Has been reported in man
Dist: AETH; OR; PAL
Hab: Abortive larva in conjunctiva or nose
Hosts: 2. Adult free-living
1. Larva in horse and other Equidae
Trans: Larvae deposited round nose
Zoo stat: 1
James 1947; Beaver *et al.* 1984

DIPTERA: PHORIDAE

Megaselia scalaris

Status: 1. Facultative parasite; rare human cases
Dist: Cosmopolitan
Hab: Larva facultatively parasitic in wounds or intestine; may also enter lungs or urino–genital system
Hosts: Normally free-living
Trans: Eggs laid on wound, or ingested
Zoo stat: 2
James 1947; Beaver *et al.* 1984; Lane and Crosskey 1993

INSECTA: SIPHONAPTERA

Numerous species of fleas occasionally bite man; in the present context, most of these are best regarded as micropredators rather than as parasites.

Pulex irritans

Status: 6. Really a micropredator but conventionally known as an ectoparasite; abundant wherever suitable conditions occur
Dist: Cosmopolitan
Hab: Predominantly inquiline
Hosts: Various, including Carnivora, pigs, rodents, burrowing owl
Trans: Active attachment to skin and clothing
Zoo stat: 3
Lewis 1972; Beaver *et al.* 1984; ICZN 2001

Lewis (1972) suggests that the genus *Pulex* almost certainly originated in the Nearctic, the 'extreme vagility of *P. irritans* permitting it to be transported throughout the world by human agencies'. Hopla (1980), on the other hand, supposes it was disseminated to the Old World on mammals, via the Beringian bridge, in the Pleistocene. He believes this despite the apparent absence of *P. irritans* from Alaskan carnivores today. Personal observations of *P. irritans* from extremely isolated Papua New Guinean villages support the early dissemination of the species.

The closest relative is *P. simulans* from western North and Central America and Hawaii (introduced), which feeds on carnivores, deer, large rodents and humans.

Tunga penetrans

Status: 5. Widely distributed and sometimes abundant; prevalence reaches 20% in Trinidad
Dist: AETH—(Introduced); NEO; OR—(Introduced)
Hab: Adult female embedded in skin
Hosts: Many large wild and domestic animals, especially pigs
Trans: Active invasion of skin
Zoo stat: 3
Lewis 1972; Beaver *et al.* 1984; Chadee 1998

According to Lewis (1972), the genus *Tunga* has a disjunct natural distribution in the Nearctic, Neotropical and eastern Palaearctic; African populations of *T. penetrans* should be seen as introductions. Hosts of other members of the genus are mainly rodents. Imported to Africa in the 17th century, but failed to establish until re-introduced in 1873 since when it has spread throughout Africa, and to the Indian sub-continent (Chadee 1998).

Exclusions

The following forms are mentioned in one or other of the major references, but have not been included, for the reason given, in the checklist.

PROTOZOA: METAMONADA

Retortamonas sinensis

Status: The single record might have been a *Chilomastix*

Kulda and Nohynkova 1978

Trichomitus faecalis

Status: The single record of this parasite, which is similar to *T. batrachorum* of amphibia, is too confused to be reliable

Honigberg 1978

PROTOZOA: SPOROZOA

Entopolypoides sp.

Syn: Probably synonymous with *Babesia microti*

Beaver *et al.* 1984; Telford and Spielman 1998

Isospora natalensis

Status: The single record was probably a pseudo-infection of *I. ohioensis* of the dog, or *I. suis* of the pig

Lindsay and Todd 1993

PROTOZOA: CILIATA

Nyctotherus faba

Status: A single case, in 1899; validity doubtful

Beaver *et al.* 1984

PLATYHELMINTHES: TURBELLARIA

Bipalium fuscatum

Status: Pseudoparasite

Coombs and Crompton 1991

Bipalium kewense

Status: Pseudoparasite

Coombs and Crompton 1991

Bipalium venosum

Status: Pseudoparasite

Coombs and Crompton 1991

TREMATODA: SCHISTOSOMATIDAE

Schistosoma incognitum

Status: The single record, from 1926, could have been in pig faeces

Muller 2000

Schistosoma margrebowei

Status: This is probably the infection reported as *S. japonicum* in southern Africa, but the identification is uncertain

Beaver *et al.* 1984

TREMATODA: PSILOSTOMIDAE

Psilorchis hominis

Status: Validity of species questionable
Beaver *et al.* 1984; Coombs and Crompton 1991

TREMATODA: HETEROPHYIDAE

Phagicola ornamenta

Status: Uncertain identification
Muller 2000

Phagicola ornata

Status: Uncertain identification
Muller 2000

Procerovum varium

Status: Experimental infection only
Coombs and Crompton 1991

TREMATODA: OPISTHORCHIDAE

Opisthorchis noverca

Status: Doubtful record, 1876

TREMATODA: MICROPHALLIDAE

Microphallus minus

Status: Experimental infection only
Coombs and Crompton 1991

TREMATODA: LECITHODENDRIIDAE

Phaneropsolus glandulosum

Poorly defined species
Muller 2002

Phaneropsolus obtusum

Poorly defined species
Muller 2002

Phaneropsolus spinicirrus

Poorly defined species
Muller 2002

TREMATODA: PARAGONIMIDAE

Paragonimus bangkokensis

Status: Human infection unknown
Blair *et al.* 1999

Paragonimus caliensis

Status: Not known as a human parasite
Blair *et al.* 1999

Paragonimus ecuadorensis

Syn: Presumably this name has been synonymised
Beaver *et al.* 1984

CESTODA: TENTACULARIIDAE

Nybelinia surmenicola

Status: Apparently mere attachment of a specimen to a tonsil: pseudoparasite
Coombs and Crompton 1991

CESTODA: DAVAINEIDAE

Raillietina asiatica

Status: A worm of doubtful identity
Beaver *et al.* 1984; Coombs and Crompton 1991

Raillietina garrisoni

Status: More detailed study required to determine taxonomic status
Beaver *et al.* 1984

Raillietina siriraji

Status: More detailed study required to determine taxonomic status
Beaver *et al.* 1984

NEMATODA: CEPHALOBIDAE

Turbatrix aceti

Status: Actively introduced with vaginal douche: pseudoparasite
Coombs and Crompton 1991

NEMATODA: RHABDITIDAE

Cheilobus quadrilabiatus

Status: Doubtful tale; pseudoparasite?
Coombs and Crompton 1991

Diploscapter coronata

Status: Doubtful tale; pseudoparasite?
Coombs and Crompton 1991

Pelodera teres

Status: Pseudoparasite
Beaver *et al.* 1984; Coombs and Crompton 1991

Rhabditis axei

Status: Identity and status uncertain
Coombs and Crompton 1991

Rhabditis niellyi

Status: Old and doubtful record
Coombs and Crompton 1991

Rhabditis pellio

Status: Probably actively introduced pseudoparasite
Coombs and Crompton 1991

Rhabditis taurica

Status: Pseudoparasite?
Coombs and Crompton 1991

Rhabditis terricola

Status: Pseudoparasite
Coombs and Crompton 1991

NEMATODA: STRONGYLOIDIDAE

Strongyloides canis

Status: Experimental infection only
Coombs and Crompton 1991

Strongyloides cebus

Status: Experimental infection only
Coombs and Crompton 1991

Strongyloides felis

Status: Experimental infection only
Coombs and Crompton 1991

Strongyloides myopotami

Status: Experimental infection only
Coombs and Crompton 1991

Strongyloides planiceps

Status: Experimental infection only
Coombs and Crompton 1991

Strongyloides procyonis

Status: Experimental infection only
Coombs and Crompton 1991

Strongyloides simiae

Status: Experimental infection only
Coombs and Crompton 1991

NEMATODA: ANCYLOSTOMATIDAE

Ancylostoma japonica

Status: Doubtfully valid species
Coombs and Crompton 1991

Ancylostoma tubaeforme

Status: Unconfirmed identification
Coombs and Crompton 1991

Bunostomum phlebotomum

Status: 'Known to be capable of causing creeping eruption' (so presumably only known experimentally)
Beaver *et al.* 1984; Coombs and Crompton 1991

Necator argentinus

Status: Doubtfully valid species
Coombs and Crompton 1991

Necator suillus

Status: Experimental infection only
Beaver *et al.* 1984; Coombs and Crompton 1991

Uncinaria stenocephala

Status: Experimental infection only
Coombs and Crompton 1991; Prociv 1998

NEMATODA: TRICHOSTRONGYLIDAE

Mecistocirrus digitatus

Status: The source of the textbook statement 'rare [in humans] in eastern countries and Central America' is unclear.

Soulsby 1982; Coombs and Crompton 1991

Nematodirus abnormalis

Status: The original record of this species as a human parasite was, according to informed sources, unreliably identified

NEMATODA: ANGIOSTRONGYLIDAE

Parastrongylus mackerrassae

Status: Infection unconfirmed

Coombs and Crompton 1991

Parastrongylus malaysiensis

Status: Infection unconfirmed

Coombs and Crompton 1991

NEMATODA: ASCARIDIDAE

Toxascaris leonina

Status: Unknown as a human parasite

Beaver *et al.* 1984; Coombs and Crompton 1991

Toxocara pteropodis

Status: Unconfirmed infection

Coombs and Crompton 1991; P. Prociv, pers. comm.

Toxocara vitulorum

Status: Unconfirmed infection

Coombs and Crompton 1991

NEMATODA: PHILOMETRIDAE

Philometra sp.

Status: Ephemeral invasion of an open wound: pseudoparasite

Coombs and Crompton 1991

NEMATODA: MERMITHIDAE

Agamomermis hominis oris

Status: Pseudoparasite

Coombs and Crompton 1991

Agamomermis restiformis

Status: Pseudoparasite

Coombs and Crompton 1991

Mermis nigrescens

Status: Pseudoparasite

Coombs and Crompton 1991

NEMATOMORPHA: GORDIIDAE

Gordius aquaticus
Status: Pseudoparasite
Coombs and Crompton 1991

Gordius chilensis
Status: Pseudoparasite
Coombs and Crompton 1991

Gordius gesneri
Status: Pseudoparasite
Coombs and Crompton 1991

Gordius inesae
Status: Pseudoparasite
Coombs and Crompton 1991

Gordius ogatai
Status: Pseudoparasite
Coombs and Crompton 1991

Gordius perronciti
Status: Pseudoparasite
Coombs and Crompton 1991

Gordius reddyi
Status: Pseudoparasite
Coombs and Crompton 1991

Gordius robustus
Status: Pseudoparasite
Coombs and Crompton 1991

Gordius setiger
Status: Pseudoparasite
Coombs and Crompton 1991

Gordius skorikowi
Status: Pseudoparasite
Coombs and Crompton 1991

NEMATOMORPHA: CHORDODIDAE

Chordodes capensis
Status: Pseudoparasite
Coombs and Crompton 1991

Neochordodes colombianus
Status: Pseudoparasite
Coombs and Crompton 1991

Parachordodes alpestris
Status: Pseudoparasite
Coombs and Crompton 1991

Parachordodes pustulosus
Status: Pseudoparasite
Coombs and Crompton 1991

Parachordodes raphaelis
Status: Pseudoparasite
Coombs and Crompton 1991

Parachordodes tolosanus
Status: Pseudoparasite
Coombs and Crompton 1991

Parachordodes violaceus
Status: Pseudoparasite
Coombs and Crompton 1991

Parachordodes wolterstorffii
Status: Pseudoparasite
Coombs and Crompton 1991

Paragordius areolatus

Status: Pseudoparasite
Coombs and Crompton 1991

Paragordius cinctus

Status: Pseudoparasite
Coombs and Crompton 1991

Paragordius esavianus

Status: Pseudoparasite
Coombs and Crompton 1991

Paragordius tricuspidatus

Status: Pseudoparasite
Coombs and Crompton 1991

Paragordius varius

Status: Pseudoparasite
Coombs and Crompton 1991

Pseudogordius tanganyikae

Status: Pseudoparasite
Coombs and Crompton 1991

INSECTA: DIPTERA

Gasterophilus nasalis

Status: Human infection unconfirmed
James 1947; Beaver *et al.* 1984

Phaenicia cuprina

Syn: *Lucilia cuprina*
Status: Semiparasite; in man, only infests wounds
James 1947; Beaver *et al.* 1984

Phormia regina

Black blowfly
Status: Semiparasite; in man, only known from wounds
James 1947; Beaver *et al.* 1984

Sarcophaga fuscicauda

Syn: *Sarcophaga peregrina*
Status: Semiparasite, only infesting necrotic tissue
James 1947; Beaver *et al.* 1984

Summary

Details of 434 species plus three distinctive subspecies are given in the checklist. These 437 forms are divided among major taxonomic groups as shown in Tables 2 and 3.

TABLE 2

Composition of the human parasite fauna by taxonomic group and abundance

	Abundance status						
Taxonomic group	*1*	*2*	*3*	*4*	*5*	*6*	*1–6*
Protozoa	38	10	11	6	7	11	**83**
Trematoda	95	8	21	4	2	0	**130**
Cestoda	38	11	2	3	0	0	**54**
Nematoda	72	9	16	6	6	5	**114**
Acanthocephala	7	0	0	0	0	0	**7**
Arthropoda	11	22	7	1	1	7	**49**
Any	**261**	**60**	**57**	**20**	**16**	**23**	**437**

TABLE 3

Composition of the human parasite fauna by taxonomic group and zoonotic status

	Zoonotic status				
Taxonomic group	*1*	*2*	*3*	*4*	*1–4*
Protozoa	23	30	20	10	**83**
Trematoda	29	93	5	3	**130**
Cestoda	12	36	2	4	**54**
Nematoda	52	40	8	14	**114**
Acanthocephala	3	4	0	0	**7**
Arthropoda	20	20	3	6	**49**
Any	**139**	**223**	**36**	**38**	**437**

The breakdown according to 'status' and 'zoonotic status' (Table 4) illustrates the large proportion of our parasites that are both purely zoonotic, and scarce or rare. Of the 437, no fewer than 362 (83%) are zoonotic (Zoo stat=1 or 2) and of these only 45 are classified as 'common', even in restricted areas (i.e. with an abundance status of at least 3). By contrast, all but six of the 75 anthroponotic or shared forms (Zoo stat=3 or 4) (17% of the total), are common at least in restricted areas. Of these 75 forms that apparently can be maintained indefinitely in humans, 38 are shared with other maintenance hosts (Zoo stat=3), leaving just 37 forms (8% of the total), which appear to depend on man alone (Zoo stat=4). The unique status of *Toxoplasma gondii,* being both abundant worldwide and purely zoonotic (man–man transmission is restricted to a single generation), is highlighted by Table 4.

TABLE 4

Composition of the human parasite fauna by abundance and zoonotic status

	Zoonotic status				
Abundance status	*1*	*2*	*3*	*4*	*1–4*
1	105	156	0	0	**261**
2	22	32	4	2	**60**
3	11	33	7	6	**57**
4	0	2	13	5	**20**
5	0	0	4	12	**16**
6	1*	0	10	12	**23**
1–6	**139**	**223**	**38**	**37**	**437**

**Toxoplasma gondii.*

Increasing information will surely alter these figures. The number of rare zoonoses must increase with new observations. The increase will be restricted as old information is seen to be unreliable, or as taxonomic revisions identify further synonymies. Taxonomic revisions, and the quantification of gene flow between parasites of humans and those of other animals, will also change the status of some of the shared and anthroponotic forms.

References

Adler, A.I. and Brancato, F.P. (1995). Human furuncular myiasis caused by *Hermetia illucens* (Diptera: Stratiomyidae). *Journal of Medical Entomology* **5**: 745–746

Andersen, K.I., Ching, H.L. and Vik, R. (1987). A review of the freshwater species of *Diphyllobothrium* with redescriptions and distribution of *D. dendriticum* and *D. ditremum* from North America. *Canadian Journal of Zoology* **65**: 2216–2228.

Anderson, R.C. (1992). *Nematode Parasites of Vertebrates*. Wallingford: CAB International, 578 pp.

Anderson, R.C., Linder, K.E. and Peregrine, A.S. (1998). *Halicephalobus gingivalis* (Stephanski, 1954) from a fatal infection in a horse in Ontario, Canada with comments on the validity of *H. deletrix* and a review of the genus. *Parasite* **5**: 255–261.

Anderson, T.J.C. (1995). *Ascaris* infections in humans from North America: molecular evidence for cross infection. *Parasitology* **110**: 215–219.

Anderson, T.J.C. and Jaenike, J. (1997). Host specificity, evolutionary relationships and macrogeographic differentiation among *Ascaris* populations from humans and pigs. *Parasitology* **115**: 325–342.

Andreassen, J. (1998). Intestinal tapeworms. In: *Topley & Wilson's Microbiology and Microbial Infections, 9th edition, Vol 5: Parasitology*, eds: F.E.G. Cox, J.P. Kreier and D. Wakelin. London: Arnold, pp 521–537.

Ashford, R.W. (1979). Occurrence of an undescribed coccidian in man in Papua New Guinea. *Annals of Tropical Medicine and Parasitology* **73**: 497–500.

Ashford, R.W. (1991). The human parasite fauna: towards an analysis and interpretation. *Annals of Tropical Medicine and Parasitology* **85**: 189–198.

Ashford, R.W. (1997). The leishmaniases as model zoonoses. *Annals of Tropical Medicine and Parasitology* **91**: 693–701.

Ashford, R.W. (1998). The leishmaniases. In: *Zoonoses*, eds: S.R. Palmer, Lord Soulsby and D.I.H. Simpson. Oxford: Oxford University Press, pp 527–544.

Ashford, R.W. (2000). Parasites as indicators of human biology and evolution. *Journal of Medical Microbiology* **49**: 771–772.

Ashford, R.W., Barnish, G. and Viney, M. (1992). *Strongyloides fuelleborni kellyi*: infection and disease in Papua New Guinea. *Parasitology Today* **8**: 314–318.

Awad El Kariem, F.M., Robinson, H.A., Dyson, D.A., Evans, D., Wright, S., Fox, M.T. and McDonald, V. (1995). Differentiation between human and animal strains of *Cryptosporidium parvum* using isoenzyme typing. *Parasitology* **110**: 129–132.

Ayala, F.J., Escalante, A.A. and Rich, S.M. (1999). Evolution of *Plasmodium* and the recent origin of world populations of *Plasmodium falciparum*. *Parassitologia* **41**: 55–68.

Bain, O., Moisson, P., Huerre, M., Landsoud-Soukate, J. and Tutin, C. (1995). Filariae from a wild gorilla in Gabon with description of a new species of *Mansonella*. *Parasite* **2**: 315–322.

Bandyopadhyay, A.K., Maji, A.K., Manna, B., Bera, D.K., Addy, M. and Nandy, A. (1995). Pathogenicity of *Artyfechinostomum oraoni* in naturally infected pigs. *Tropical Medicine and Parasitology* **46**: 138–139.

Basset, D., Girou, C., Nozais, I.P., d'Hermies, F., Hoang, C., Gordon, R. and d'Alessandro, A. (1998). Neotropical echinococcosis in Suriname. *Echinococcus oligarthus* in the orbit and *Echinococcus vogeli* in the abdomen. *American Journal of Tropical Medicine and Hygiene* **59**: 787–790.

Beck, A.M. (2000). The human-dog relationship: a tale of two species. In: *Dogs, Zoonoses and Public Health*, eds: C.N.L. Macpherson, F.X. Meslin and A. Wandeler. Wallingford: CAB International, pp 1–16.

Beaver, P.C., Jung, R.C. and Cupp, E.W. (1984). *Clinical Parasitology, 9th edition*. Philadelphia: Lea and Febiger, 825 pp.

Beaver, P.C., Wolfson, J.J., Waldon, M.A., Swartz, M.N., Evans, G.W. and Adler, J. (1987). *Dirofilaria ursi*-like parasites acquired in northern United States and Canada: report of two cases and brief review. *American Journal of Tropical Medicine and Hygiene* **37**: 357–362.

Beesley, W.N. (1998). Scabies and other mite infestations. In: *Zoonoses*, eds: S.R. Palmer, Lord Soulsby and D.I.H. Simpson. Oxford: Oxford University Press, pp 859–872.

Biocca, E. (1960). Osservazioni sulla morfologia e biologia del ceppo sardo di *Schistosoma bovis* e sulla dermatite umana da esso provocata. *Parassitologia* **2**: 47–54.

Blair, D., Xu, Z.-B. and Agatsuma, T. (1999). Paragonimiasis and the genus *Paragonimus*. *Advances in Parasitology* **42**:113–222.

Boreham, P.F.L. and Stenzel, D.J. (1998). Blastocystosis. In: *Zoonoses*, eds: S.R. Palmer, Lord Soulsby and D.I.H. Simpson. Oxford: Oxford University Press, pp 625–634.

Boussinesq, M. and Gardon, J. (1997). Prevalences of *Loa loa* microfilaraemia throughout the area endemic for the infection. *Annals of Tropical Medicine and Parasitology* **91**: 573–589.

Boussinesq, M., Bain, O., Chabaud, A.G., Gardon-Wendel, N., Kamgno, J. and Chippaux, J.P. (1995). A new zoonosis of the cerebrospinal fluid of man probably caused by *Meningonema peruzzi*, a filaria of the central nervous system of Cercopithecidae. *Parasite* **2**: 173–176.

Butcher, A.R. and Grove, D.I. (2001). Description of the life-cycle stages of *Brachylaima cribbi* n. sp. (Digenea: Brachylaimidae) derived from eggs recovered from human faeces in Australia. *Systematic Parasitology* **49**: 211–221.

Cacciapuoti, R. (1947). Su di una nuova distomatosi umana in Ethiopia. *Rivista di Biologia Coloniale* **8**: 111–116.

Caccio, S., Pinter, E., Fantini, R., Mezzaroma, I. and Pozio, E. (2002). Human infection with *Cryptosporidium felis*: case report and literature review. *Emerging Infectious Diseases* **8**: 85–86.

Campbell, W.C. (1991). *Trichinella* in Africa and the *nelsoni* affair. In: *Parasitic Helminths and Zoonoses in Africa*, eds: C.M.L. Macpherson and P.S. Craig. London: Unwin Hyman, pp 83–100.

Campos, D.M.B., Santos, E.R., Paco, J.M. and Souza, M.A. (1995). Lagochilascariase humana. Registro de um novo caso procedente do sul do Para. *Revista de Patologia Tropical* **24**: 313–322.

Canning, E.U. (1998). Microsporidiosis. In: *Zoonoses*, eds: S.R. Palmer, Lord Soulsby and D.I.H. Simpson. Oxford: Oxford University Press, pp 609–623.

Canning, E.U. (2001). Microsporidia. In: *Principles and Practice of Clinical Parasitology*, eds: S. Gillespie and R.D. Pearson. London: John Wiley & Sons, pp 171–195.

Chadee, D.D. (1998). Tungiasis among five communities in south-western Trinidad, West Indies. *Annals of Tropical Medicine and Parasitology* **92**: 107–113.

Chai, J. Y., Han, E.T., Park, Y. K., Guk, S. M. and Lee, S. H. (2001). *Acanthoparyphium tyosenense*: the discovery of human infection and identification of its source. *Journal of Parasitology* **87**: 794–800.

Chai, J.Y., Song, T.E., Han, E.T., Guk, S.M., Park, Y.K., Choi M.H. and Lee, S.H. (1998). Two endemic foci of heterophyids and other intestinal fluke infections in southern and western coastal areas in Korea. *Korean Journal of Parasitology* **36**: 155–161.

Chai, J.Y. and Lee, S.H. (2002). Food-borne intestinal trematode infections in the republic of Korea. *Parasitology International* **51**: 129–154.

Cheng, T.C. (1998). Anisakiosis. In: *Zoonoses*, eds: S.R. Palmer, Lord Soulsby and D.I.H. Simpson. Oxford: Oxford University Press, pp 823–840.

Chittenden, A.M. and Ashford, R.W. (1987). *Enterobius gregorii* Hugot 1983: first report in UK. *Annals of Tropical Medicine and Parasitology* **81**: 195–198.

Chowdhury, N. and Tada, I. (2001). *Perspectives on Helminthology*. Enfield (NH): Science Publishers, 531 pp.

Coatney, G.R., Collins, W.E., Warren, McW. and Contacos, P.G. (1971). *The Primate Malarias*. Bethesda: National Institutes of Health, 366 pp.

Coombs, I. and Crompton, D.W.T. (1991). *A Guide to Human Helminths*. London: Taylor and Francis, 196 pp.

Cox, F.E.G. (1998). Babesiosis and malaria. In: *Zoonoses*, eds: S.R. Palmer, Lord Soulsby and D.I.H. Simpson. Oxford: Oxford University Press, pp 599–608.

Cox, F.E.G., Kreier, J.P. and Wakelin, D. (eds) (1998). *Topley & Wilson's Microbiology and Microbial Infections, 9th edition, Vol 5: Parasitology*. London: Arnold, 701 pp.

Cross, J.H. (1998a). Capillariosis. In: *Zoonoses*, eds: S.R. Palmer, Lord Soulsby and D.I.H. Simpson. Oxford: Oxford University Press, pp 759–772.

Cross, J.H. (1998b). Angiostrongylosis. In: *Zoonoses*, eds: S.R. Palmer, Lord Soulsby and D.I.H. Simpson. Oxford: Oxford University Press, pp 773–782.

Curry, A. (1998). Microsporidians. In: *Topley & Wilson's Microbiology and Microbial Infections, 9th edition, Vol 5: Parasitology*, eds: F.E.G. Cox, J.P. Kreier and D. Wakelin. London: Arnold, pp 411–430.

Curry, A. (1999). Human microsporidial infection and possible animal sources. *Current Opinion in Infectious Diseases* **12**: 473–480.

Dedet, J.-P. (1993). *Leishmania* et leishmanioses du continent americain. *Annales de l'Institut Pasteur/Actualités* **4**: 3–25.

Dedet, J.-P., Roche, B., Pratlong, F., Cales-Quist, D., Jouanelle, J., Benichou, J.C. and Huerre, M. (1995). Diffuse cutaneous infection caused by a presumed monoxenous trypanosomatid in a patient infected with HIV. *Transactions of the Royal Society of Tropical Medicine and Hygiene* **89**: 644–646.

Deluol, A.M. and Cenac, J. (1994). Les microsporidioses *Microsporidiosis*. *Annales de Biologie Cininque* **52**: 37–44.

Delyamure, S.L. (1968). *Helminthofauna of Marine Mammals* (translated by M. Raveh). Jerusalem: Israel Programme for Scientific Translations, 522 pp.

Denegri, G.M. and Perez-Serrano, J. (1997). Bertiellosis in man: a review of cases. *Revista do Instituto de Medicina Tropical de São Paulo* **39**: 123–127.

Denham, D.A. (1998). Zoonotic infections with filarial nematodes. In: *Zoonoses*, eds: S.R. Palmer, Lord Soulsby and D.I.H. Simpson. Oxford: Oxford University Press, pp 783–788.

Dennett, X., Siejka, S., Andrews, J.R.H., Beveridge, I. and Spratt, D.M. (1998). Polymyositis caused by a new species of nematode. *Medical Journal of Australia* **168**: 226–227.

Desportes-Livage, I. (1996). Human microsporidiosis and AIDS: recent advances. *Parasite* **3**: 107–113.

Diamond, L.S. and Clark, G. (1993). A redescription of *Entamoeba histolytica* separating it from *E. dispar*. *Journal of Eukaryotic Microbiology* **40**: 340–344.

Didier, E.S., Snowden, K.F. and Shadduck, A. (1998). Biology of microsporidian species infecting mammals. *Advances in Parasitology* **40**: 283–320.

Dissanaike, A.S., Bandara, C.D., Padmini, H.H., Thalamulla, R.L. and Naotunne, T.S. (2000). Recovery of a species of *Brugia*, probably *Brugia ceylonensis* from the conjunctiva of a patient in Sri Lanka. *Annals of Tropical Medicine and Parasitology* **94**: 83–86.

Dissanaike, A.S., Hock, Q.C. and Min, T.S. (1974). Mature female filaria, probably *Brugia* sp., from the conjunctiva of a man in west Malaysia. *American Journal of Tropical Medicine and Hygiene* **23**: 1023–1026.

Ditrich, O., Palkovic, L., Sterba, J., Prokopic, J., Loudova, J. and Giboda, M. (1991). The first finding of *Cryptosporidium baileyi* in man. *Parasitology Research* **77**: 44–47.

Dubey, J.-P., Speer, C.A. and Fayer, R. (1989). *Sarcocystosis of Animals and Man*. Boca Raton: CRC Press, 215 pp.

Eckert, J. (1998). Alveolar echinococcosis and other forms of echinococcosis. In: *Zoonoses*, eds: S.R. Palmer, Lord Soulsby and D.I.H. Simpson. Oxford: Oxford University Press, pp 689–716.

Eom, K.S. and Rim, H.-J. (1993). Morphologic descriptions of *Taenia asiatica* sp. n. *Korean Journal of Parasitology* **31**: 1–6.

Escalante, A.A., Barrio, E. and Ayala, F.J. (1995). Evolutionary origin of human and primate malarias: evidence from the circumsporozoite protein gene. *Molecular Biology and Evolution* **12**: 616–626.

Fayer, R., Trout, J.M., Xiao, L., Lai, A.A. and Dubey, J.P. (2001). *Cryptosporidium canis* n. sp. from domestic dogs. *Journal of Parasitology* **87**: 1415–1422.

Francois, A., Favennec, L., Cambon-Michot, C., Gueit, I., Biga, N., Tron, F., Brasseur, P. and Hemet, J. (1998). *Taenia crassiceps* invasive cysticercosis: a new human pathogen in acquired immunodeficiency syndrome? *American Journal of Surgical Pathology* **22**: 488–492.

Freitas, A.L., de Carli, G. and Blankenhein, M.H. (1995). *Mammomonogamus (Syngamus) laryngaeus* infection: a new Brazilian human case. *Revista do Instituto de Medicina Tropical de São Paulo* **37**: 177–179.

Gatei, W., Ashford R.W., Beeching, N.J., Kamwathi, S.K., Greenshill, J. and Hart, C.A. (2002). *Cryptosporidium muris* infection in an HIV-infected adult, Kenya. *Emerging Infectious Diseases* **8**: 204–206.

Goldsmid, J.M. (1991). The African hookworm problem: an overview. In: *Parasitic Helminths and Zoonoses in Africa*, eds: C.M.L. Macpherson and P.S. Craig. London: Unwin Hyman, pp 101–137.

Hall, M. and Wall, R. (1995). Myiasis of humans and domestic animals. *Advances in Parasitology* **35**: 258–334.

Harinasuta, K.T., Bunnag, D. and Radomyos, R.B. (1987). Intestinal fluke infections. In: *Intestinal Helminthic Infections: Baillère's Clinical Tropical Medicine and Communicable Diseases, Vol 2, No 3*, ed: Z.S. Pawlowski. London: Baillière Tindall, pp 695–721.

Harrison, L.J.S. and Sewell, M.M.H. (1991). The zoonotic *Taeniae* of Africa. In: *Parasitic Helminths and Zoonoses in Africa*, eds: C.M.L. Macpherson and P. S. Craig. London: Unwin Hyman, pp 54–82.

Hartsheerl, R.A., Van Gool, T., Schuitena, A.R.J., Didier, E.S. and Terpstra, W.J. (1995). Genetic and immunological characterisation of the microsporidium *Septata intestinalis*: reclassification to *Encephalitozoon intestinalis*. *Parasitology* **110**: 277–285.

Hasegawa, H., Takao, Y., Nakao, M., Fukuma, O. and Ide, K. (1998). Is *Enterobius gregorii* Hugot, 1983 (Nematoda: Oxyuridae) a distinct species? *Journal of Parasitology* **54**: 131–134.

Haque, R., Ali, I.K.M., Clark, G. and Petri, A. (1998). A case report of *Entamoeba moshkovskii* infection in a Bangladeshi child. *Parasitology International* 47: 201–202.

Hawdon, J.M. and Johnston, S.A. (1996). Hookworms in the Americas: an alternative to trans-Pacific contact. *Parasitology Today* **12**: 72–74.

Hoberg, F.P., Alkire, N.L., de Querioz, A. and Jones, A. (2000). Out of Africa: origins of the *Taenia* tapeworms in humans. *Proceedings of the Royal Society Series B* **268**: 781–787.

Hollister, W.S., Canning, E.U. and Viney, M. (1989). Prevalence of antibodies to *Encephalitozoon cuniculi* in stray dogs as determined by an ELISA. *Veterinary Record* **124**: 332–336.

Hollister, W.S., Canning, E.U., Weidner, E., Field, A.S., Kench, J. and Marriot, D.J. (1996). Development and ultrastructure of *Trachipleistophora hominis* n.g., n.sp. after isolation from an AIDS patient and inoculation into athymic mice. *Parasitology* **112**: 143–154.

Honigberg, B.M. (1978). Trichomonads of importance in human medicine. In: *Parasitic Protozoa Vol 2*, ed: J.P. Kreier. New York: Academic Press, pp 275–454.

Hopla, C.E. (1980). A study of the host associations and zoogeography of *Pulex*. In: *Fleas*, eds: R. Traub and H. Starcke. Amsterdam: A.A. Balkema, pp 185–207.

Huffman, J.E. and Fried, B. (1990). *Echinostoma* and echinostomiasis. *Advances in Parasitology* **29**: 215–269.

Hugot, J.P. (1993). Redescription of *Enterobius anthropopitheci* (Gedoelst 1916) (Nematoda: Oxyuridae), a parasite of chimpanzees. *Systematic Parasitology* **26**: 201–207.

International Commission for Zoological Nomenclature (2001). *Official Lists and Indexes of Names and Works in Zoology, Supplement 1986–2000*. London: International Trust for Zoological Nomenclature, 136 pp.

James, M.T. (1947). *The Flies that Cause Myiasis in Man. Miscellaneous Publication No 631*. Washington: United States Department of Agriculture, 175 pp.

John, D.T. (1998). Opportunistic amoebae. In: *Topley & Wilson's Microbiology and Microbial Infections, 9th edition, Vol 5: Parasitology*, eds: F.E.G. Cox, J.P. Kreier and D. Wakelin. London: Arnold, pp 179–192.

Jongwutiwes, S., Chantachum, N., Kraivichian, P., Siryasatien, P., Putaporntip, C., Tamburrini, A., La Rosa, G., Sreesunpasirikul, C., Yingyourd, P. and Pozio, E. (1998). First outbreak of human trichinellosis caused by *Trichinella pseudospiralis*. *Clinical Infectious Diseases* **26**: 111–115.

Jordan, P., Webbe, G. and Sturrock, R.F. (1993). *Human Schistosomiasis*. Wallingford: CAB International, 465 pp.

Keith, A.C. (1999). Three incidents of human myiasis by rodent *Cuterebra* (Diptera: Cuterebridae) larvae in a localized region of Western Pennsylvania. *Journal of Medical Entomology* **36**: 831–832.

Khalil, L.F. (1991). Zoonotic helminths of wild and domestic animals in Africa. In: *Parasitic Helminths and Zoonoses in Africa*, eds: C.M.L. Macpherson and P. S. Craig. London: Unwin Hyman, pp 260–272.

Kino, H., Hori, W., Kobayashi, H., Nakamura, N. and Nagasawa, K. (2002). A mass occurrence of human infection with *Diplogonoporus grandis* (Cestoda: Diphyllobothriidae) in Shizuoka Prefecture, central Japan. *Parasitology International* **51**: 73–79.

Komba, E.K., Kibona, S.N., Ambwena, A.K., Stevens, J.R. and Gibson, W.C. (1998). Genetic diversity among *Trypanosoma b. brucei* isolates from Tanzania. *Parasitology* **115**: 571–579.

Kulda, J. and Nohynkova, E. (1978). Intestinal flagellates. In: *Parasitic Protozoa Vol 2*, ed: J.P. Kreier, J.P. New York: Academic Press, pp 1-138.

Lane, R.P. and Crosskey, R.W. (1993). *Medical Insects and Arachnids*. London: Chapman and Hall, 723 pp.

Lapierre, J. and Rousset, J.J. (1973). L'infestation à protozoaires buccaux. *Annales de Parasitologie Humaine et Comparée* **48**: 205–216.

Ledee, D.R., Hay, J., Byers, T.J., Seal, D.V. and Kirkness, C.N. (1996). *Acanthamoeba griffini*. Molecular characterization of a new corneal pathogen. *Investigative Ophthalmology and Visual Science* **37**: 544–550.

Lee, H.L., Chandrawathani, P., Wong, W.Y., Tharam, S. and Lim, W.Y. (1995). A case of enteric myiasis due to larvae of *Heremetia illucens* (Family: Stratiomyidae): first report in Malaysia. *Malaysian Journal of Pathology* **17**: 109–111.

Lee, S.H. and Chai, J.Y. (2001). A review of *Gymnophalloides seoi* (Digenea: Gymnophallidae) and human infections in the Republic of Korea. *Korean Journal of Parasitology* **39**: 85–118.

Levine, N.D. (1985). *Veterinary Protozoology*. Ames: Iowa State University Press, 414 pp.

Lewis, R.E. (1972). Notes on the geographical distribution and host preferences in the order Siphonaptera. *Journal of Medical Entomology* **9**: 511–520.

Lindsay, D.S. and Todd, K.S. (1993). Coccidia of mammals. In *Parasitic Protozoa, 2nd edition, Vol 4,* ed: J.P. Kreier. San Diego: Academic Press, pp 89–131.

Lloyd, S. (1998a). Other cestode infections: hymenolepiosis, diphyllobothriosis, coenurosis and other larval and adult cestodes. In: *Zoonoses*, eds: S.R. Palmer, Lord Soulsby and D.I.H. Simpson. Oxford: Oxford University Press, pp 635–649.

Lloyd, S. (1998b). Toxocarosis. In: *Zoonoses*, eds: S.R. Palmer, Lord Soulsby and D.I.H. Simpson. Oxford: Oxford University Press, pp 841–854.

Lloyd, S. and Soulsby, E.J.L. (1998). Other trematode infections. In: *Zoonoses*, eds: S.R. Palmer, Lord Soulsby and D.I.H. Simpson. Oxford: Oxford University Press, pp 731–746.

Lowman, P.M., Takvorian, P.M. and Cali, M. (2000). The effects of elevated temperatures and various time-temperature combinations on the development of *Brachiola* (*Nosema*) *algerae* n. comb. in mammalian cell culture. *Journal of Eukaryote Microbiology* **47**: 221–234.

Lumbreras-Cruz, H., Terashima Iwashita, A., Alvarez-Bianchi, H. and Casanova, R. (1986). Variacion en la incidencia del parasitismo por *Diphyllobothrium pacificum* en 20 anos. In: *Abstracts of the 25th Anniversary Scientific Days of the Universidad Peruana Cayetana Heridia, Lima, Peru, 8–9 Sept 1986*, Lima: Universidad Peruana Cayetana Heridia, p 295.

Maji, A.K., Bera, D.K., Manna, B., Nandy, A., Addy, M and Bandyopadhyay, A.K. (1993). First record of human infection with *Echinostoma malayanum* in India. *Transactions of the Royal Society of Tropical Medicine and Hygiene* **87**: 673.

Malek, E.A. (1980a). *Snail-transmitted Parasitic Diseases, Vol 1.* Baton Roca: CRC Press, 334 pp.

Malek, E.A. (1980b). *Snail-transmitted Parasitic Diseases, Vol 2.* Baton Roca: CRC Press, 324 pp.

Martinez-Palomo, A. and Cantellano, M.E. (1998). Intestinal amoebae. In: *Topley & Wilson's Microbiology and Microbial Infections, 9th edition, Vol 5: Parasitology*, eds: F.E.G. Cox, J.P. Kreier and D. Wakelin. London: Arnold, pp 157–177.

McCarthy, J. and Moore, T.A. (2000). Emerging helminth zoonoses. *International Journal for Parasitology* **30**: 1351–1360.

Moncada, L.I., Lopez, M.C., Murcia, M.I., Nicholls, S., Leon, F., Guio, O.L. and Corredor, A. (2001). *Myxobolus* sp., another opportunistic parasite in immunosuppressed patients? *Journal of Clinical Microbiology* **39**: 1938–1940.

Moravec, F. (2001). Redescription and systematic status of *Capillaria philippinensis*, an intestinal parasite of human beings. *Journal of Parasitology* **87**: 161–164.

Muller, R. (1998). Dracunculiasis. In: *Topley & Wilson's Microbiology and Microbial Infections, 9th edition, Vol 5: Parasitology*, eds: F.E.G. Cox, J.P. Kreier and D. Wakelin. London: Arnold, pp 661–665.

Muller, R. (2000). Dogs and trematode zoonoses. In: *Dogs,Zoonoses and Public Health*, eds: C.N.L. Macpherson, F.X. Meslin and A.I. Wandeler. Wallingford: CAB International, pp 149–176.

Muller, R. (2002).*Worms and Human Disease* (Second edition). Wallingford: CAB International, 300 pp.

Ning, C.X. (2001). A case with skin myiasis caused by *Gasterophilus nigricornis*, *Chinese Journal of Parasitology and Parasitic Diseases* **19**: 60. (From Annual Review of Medical and Veterinary Entomology)

Nolan, T.J. (1998). Trichostrongylosis. In: *Zoonoses*, eds: S.R. Palmer, Lord Soulsby and D.I.H. Simpson. Oxford: Oxford University Press, pp 855–858.

Nomura, Y., Nagakura, K., Kagei, N., Tsutsumi, Y., Araki, K. and Sugawara, M. (2000). Gnathostomiasis possibly caused by *Gnathostoma malaysiae*. *Tokai Journal of Experimental and Clinical Medicine* **25**: 1–6.

Noyes, H., Pratlong, F., Chance, M.L., Ellis, J., Lanotte, G. and Dedet, J.-P. (2002). A previously unclassified trypanosomatid responsible for human cutaneous lesions in Martinique (French West Indies) is the most divergent member of the genus *Leishmania ss*. *Parasitology* **124**: 1724.

Ooi, H.K., Tenora, F., Itoh, K. and Kamiya, M. (1993). Comparative study of *Trichuris trichiura* from non-human primates and from man, and their difference with *T. suis*. *Journal of Veterinary Medical Science* **55**: 363–366.

Orihel, T.C. and Eberhard, M.L. (1998). Zoonotic filariasis. *Clinical Microbiology Reviews* **11**: 366–381.

Orihel, T.C. and Isbey, E.K. (1990). *Dirofilaria striata* infection in a North Carolina child. *American Journal of Tropical Medicine and Hygiene* **42**: 124–126.

Ouma, J.H. and Fenwick, A. (1991). Animal reservoirs of schistosomiasis. In: *Parasitic Helminths and Zoonoses in Africa*, eds: C.M.L. Macpherson and P.S. Craig. London: Unwin Hyman, pp 224–236.

Palmer, S.R., Lord Soulsby and Simpson, D.I.H. (eds) (1998). *Zoonoses*. Oxford: Oxford University Press, 948 pp.

Pampiglione, S. (1958). Indagine epidemiologica sulla miasi umana da *Oestrus ovis* in Italia. Nota II. *Nuovi Annali d'Igiene e Microbiologia* **9**: 494–517.

Pampiglione, S. and Ricciardi, M.L. (1972). Geographic distribution of *Strongyloides fülleborni* in humans in tropical Africa. *Parassitologia* **14**: 329–338.

Pampiglione, S., Canestri, G. and Rivasi, F. (1995). Human dirofilariasis due to *Dirofilaria* (*Nochtiella*) *repens*: a review of world literature. *Parassitologia* **37**: 149–193.

Pampiglione, S., Vakalis, N., Lyssimachu, A., Kouppari, G. and Orihel, T.C. (2001), Subconjunctival zoonotic *Onchocerca* in an Albanian man. *Annals of Tropical Medicine and Parasitology* **95**: 827–832.

Peters, W., Garnham, P.C.C., Killick-Kendrick, R., Rajapaksa, N., Cheong, W.H. and Cadigan, F.C. (1976). Malaria parasites of the orang-utan (*Pongo pygmaeus*) in Borneo. *Proceedings of the Royal Society, Series B* **275**: 439–482.

Polderman, A.M. and Blotkamp, J. (1995). *Oesophagostomum* infection in humans. *Parasitology Today* **12**: 441–481.

Pozio, E. and La Rosa, G. (2000). *Trichinella murrelli* n.sp: etiological agent of sylvatic trichinellosis in temperate areas of North America. *Journal of Parasitology* **86**: 134–139.

Pozio, E., La Rosa, G., Murrell, K.D. and Lichtenfels, J.R. (1992). Taxonomic revision of the genus *Trichinella*. *Journal of Parasitology* **78**: 654–659.

Premvati, G.R. and Pande, V. (1974). On *Artyfechinostomum malayanum* (Leiper 1911) Mendheim 1943 (Trematoda: Echinostomatidae) with synonymy of allied species and genera. *Proceedings of the Helminthological Society of Washington* **41**: 151–160.

Prociv, P. (1998). Zoonotic hookworm infections. In: *Zoonoses*, eds: S.R. Palmer, Lord Soulsby and D.I.H. Simpson. Oxford: Oxford University Press, pp 803–822.

Qari, S.H., Shi, Y.-P., Goldman, I.F., Udhayakumar, V., Alpers, M.P., Collins, W.E. and Lal, A.A. (1993). Identification of *Plasmodium vivax*-like human parasite. *Lancet* **341**: 780–783.

Rausch, R.H. and Hilliard, D.K. (1970). Studies on the helminth fauna of Alaska. 49. The occurrence of *Diphyllobothrium latum* in Alaska with notes on other species. *Canadian Journal of Zoology* **48**: 1201–1219.

Riley, J. (1986). The biology of pentastomids. *Advances in Parasitology* **25**: 45–128.

Samet, J., Bignami, G.S., Feldman, R., Hawkins, W., Neff, J. and Smayda, T. (2001). *Pfiesteria*: review of the science and identification of research gaps. Report for the National Center for Environmental Health, Centers for Disease Control and Prevention. *Environmental Health Perspectives* **109** Supplement 5: 639–659.

Sang, D.K., Pratlong, F. and Ashford, R.W. (1992). The identity of *Leishmania tropica* in Kenya. *Transactions of the Royal Society of Tropical Medicine and Hygiene* **86**: 621–622.

Sargeaunt, P.G., Patrick, S. and O'Keefe, D. (1992). Human infections of *Entamoeba chattoni* masquerade as *Entamoeba histolytica*. *Transactions of the Royal Society of Tropical Medicine and Hygiene* **86**: 633–634.

Schad, G.A. (1991). The parasite. In: *Hookworm Infection*, eds: H.M. Gilles and P.A.J. Ball. Amsterdam: Elsevier, pp 15–49.

Schad, G.A. and Nawalinski, T.A. (1991). Historical introduction. In: *Hookworm Infection*, eds: H.M. Gilles and P.A.J. Ball. Amsterdam: Elsevier, pp 1–14.

Schaefer, C.W. (1978). Ecological separation of the human head lice and body lice. *Transactions of the Royal Society of Tropical Medicine and Hygiene* **72**: 669–670.

Schmidt, G.D. (1971). Acanthocephalan infections of man with two new records. *Journal of Parasitology* **57**: 582–584.

Singh, S., Samantaray, J.C., Singh, N., Das, G.B. and Verma, I.C. (1993). *Trichuris vulpis* infection in an Indian tribal population. *Journal of Parasitology* **79**: 457–458.

Skrjabin, K.I. (1952). *Trematodes in Animals and Man: Principles of Trematology Vol 6*. Moscow: Academy of Sciences of the USSR, 759 pp.

Smith, H.V., Paton, C.A., Girdwood, R.W.A. and Mtambo, M.M.A. (1996). *Cyclospora* in non-human primates in Gombe, Tanzania. *Veterinary Record* **May 25**: 528.

Smyth, J.D. (1995). Rare, new and emerging helminth zoonoses. *Advances in Parasitology* **36**: 1–45.

Sobhon, P. and Upatham, P. (1990). *Snail Hosts, Life-cycle and Tegumental Structure of Oriental Schistosomes*. Geneva: World Health Organization, 321 pp.

Telford, S.R. and Spielman, A. (1998). Babesiosis of humans. In: *Topley & Wilson's Microbiology and Microbial Infections, 9th edition, Vol 5: Parasitology*, eds: F.E.G. Cox, J.P. Kreier and D. Wakelin. London: Arnold, pp 349–359.

Thompson, R.C.A., Hopkins, R.M. and Homan, W.L. (2000). Nomenclature and genetic groupings of *Giardia* infecting mammals. *Parasitology Today* **16**: 210–213.

Thompson, R.C.A., Reynoldson, J.A. and Lymbery, A.J. (eds) (1994). Giardia: *from Molecules to Disease.* Wallingford: CAB International, 394 pp.

Thompson, R.C.A. and Chalmers, R.M. (2002). *Cryptosporidium*: from molecules to disease. *Trends in Parasitology* **18**: 98–100.

Truc, P., Formenty, P., Duvallet, G., Komoin-Oka, C., Diallo, P.B. and Lauginie, F. (1997). Identification of trypanosomes isolated by KIVI from wild mammals in Côte d'Ivoire: diagnostic, taxonomic and epidemiological considerations. *Acta Tropica* **67**: 187–196.

Tsuji, M., Wei, Q., Zamoto, A., Morita, C., Arai, S., Shiota, T., Fujimagari, M., Itagaki, A., Fujita, H. and Ishihara, C. (2001). Human babesiosis in Japan: epizoologic survey of rodent reservoir and isolation of new type of *Babesia microti*-like parasite. *Journal of Clinical Microbiology* **12**: 4316–4322.

Visvesvara, G.S., Belloso, M., Moura, H., DaSilva, A.J., Moura, I.N.S., Leitch, G. J., Schwartz, D.A., Chevez-Barrios, S., Pieniazek, N.J. and Goosey, J.D. (1999). Isolation of *Nosema algerae* form the cornea of an immunocompetent patient. *Journal of Eukaryote Microbiology* **46**: 10S.

Warhurst, D.C. (1985). Pathogenic free-living amoebae. *Parasitology Today* **1**: 24–28.

Weinman, D. (1977). Trypanosomiasis of man and macaques. In: *Parasitic Protozoa, Vol 1: Taxonomy, Kinetoplastids and Flagellates of Fish*, ed: J.P. Kreier. New York: Academic Press: pp 329–355.

Wong, K.T. and Pathmanathan, R. (1992). High prevalence of human sarcocystosis in south-east Asia. *Transactions of the Royal Society of Tropical Medicine and Hygiene* **86**: 631–632.

Yagita, K., Izumiyama, S., Tachibana, H., Masuda, G., Iseki, M., Furuya, K., Kameoka, Y., Kuroki, T., Itagaki, T. and Endo, T. (2001). Molecular characterization of *Cryptosporidium* isolates obtained from human and bovine infections in Japan. *Parasitology Research* **87**: 950–955.

Yamaguchi, T. (1981). *A Colour Atlas of Clinical Parasitology*. London: Wolfe, 293 pp.

Yoshida, Y. (1997). *Illustrated Human Parasitology, 5th edition.* Tokyo: Nanzando, 284 pp.

Yoshikawa, H., Yamada, M., Matsumoto, Y. and Yoshida, Y. (1989). Variations in egg size of *Trichuris trichiura*. *Parasitology Research* **75**: 649–654.

Yoshimura, K. (1998). *Angiostrongylus* and less common nematodes. In: *Topley & Wilson's Microbiology and Microbial Infections, 9th edition, Vol 5: Parasitology*, eds: F.E.G. Cox, J.P. Kreier and D. Wakelin. London: Arnold, pp 635–659.

Zaman, V. (1998). *Balantidium coli.* In: *Topley & Wilson's Microbiology and Microbial Infections, 9th edition, Vol 5: Parasitology*, eds: F.E.G. Cox, J.P. Kreier and D. Wakelin. London: Arnold, pp 445–450.

Zhang, Q., Wang, B. and Huang, M. (1996). *Armillifer agkistrodontis* disease: report of case. *Zhonghua Nei Ke Za Zhi* **35**: 747–749. (From Pub Med.)

INDEX

E

F

G

H

I

K

L

M

N

O

P

R

INDEX

S

T

U

V

W

INDEX